Iron and Steel

Modern civilization as we know it would not be possible without iron and steel. Steel is essential in the machinery necessary for the manufacture of all our needs. Even the words themselves have come to suggest strength. Phrases such as *iron willed, iron fisted, iron clad, iron curtain,* and *pumping iron* imply strength. A *steely glance* is a stern look. A *heart of steel* refers to a very hard demeanor. The Russian dictator Stalin (which means "steel" in Russian) chose the name to invoke fear in his subordinates. This book is intended both as a resource for engineers and as an introduction to the layman about our most important metal system. After an introduction that deals with the history and refining of iron and steel, the rest of the book examines their physical properties and metallurgy.

William F. Hosford is Professor Emeritus of Materials Science at the University of Michigan. He is the author of numerous research publications and the following books: *The Mechanics of Crystals and Textured Polycrystals* (1993); *Physical Metallurgy* (2005); *Materials Science: An Intermediate Text* (2007); *Materials for Engineers* (2008); *Reporting Results: A Practical Guide for Scientists and Engineers* with David C. Van Aken (2008); *Mechanical Behavior of Materials*, 2nd edition (2009); *Wilderness Canoe Tripping* (2009); *Solid Mechanics* (2010); *Physical Metallurgy*, 2nd edition (2010); and *Metal Forming: Mechanics and Metallurgy*, 4th edition, with Robert M. Caddell (2011).

IRON AND STEEL

William F. Hosford

University of Michigan

CAMBRIDGE
UNIVERSITY PRESS

CAMBRIDGE
UNIVERSITY PRESS

32 Avenue of the Americas, New York NY 10013-2473, USA

Cambridge University Press is part of the University of Cambridge.

It furthers the University's mission by disseminating knowledge in the pursuit of
education, learning and research at the highest international levels of excellence.

www.cambridge.org
Information on this title: www.cambridge.org/9781107652934

© William F. Hosford 2012

First published 2012
First paperback edition 2014

A catalogue record for this publication is available from the British Library

Library of Congress Cataloguing in Publication data
Hosford, William F.
Iron and steel / William F. Hosford.
    p.   cm.
Includes bibliographical references and index.
ISBN 978-1-107-01798-6 (hardback)
1. Iron – History.   2. Steel – History.   3. Iron – Metallurgy.
4. Steel – Metallurgy.   I. Title.
TN703.H67   2012
669'.1–dc23   2011037262

ISBN   978-1-107-01798-6 Hardback
ISBN   978-1-107-65293-4 Paperback

CONTENTS

Preface   *page* xi

1 General Introduction **1**
   Nomenclature   1
   Phases   2
   Production   3

2 Early History of Iron and Steel **4**
   Native Iron   4
   Wrought Iron   6
   Steel   8
   References   10

3 Modern Steel Making **11**
   Blast Furnace   11
   Coke   13
   Bessemer Steel-Making Process   13
   Open-Hearth Steel-Making Process   15
   Basic Oxygen Furnace   16
   Electric Arc Process   18
   Furnace Linings and Slags   19
   Casting   20
   Hot Rolling   21
   Cold Rolling   22
   Recycling   22
   References   24

4 Constitution of Carbon Steels **25**
Microstructures of Carbon Steels   25
Pearlite Formation   30
References   34

5 Plastic Strength **35**
Dislocation Density   35
Strain Hardening   35
Grain Size   37
Solute Effects   39
Temperature Dependence   39
Hardness   40
Strain-Rate Dependence of Flow Stress   40
Combined Effects of Temperature and Strain Rate   46
Superplasticity   49
Strength Differential Effect   49
References   50

6 Annealing **51**
General   51
Recovery   51
Relief of Residual Stresses   54
Recrystallization   54
Grain Growth   58
References   65

7 Deformation Mechanisms and Crystallographic Textures **66**
Slip and Twinning Systems   66
Wire Textures in bcc Metals   66
Rolling Texture   71
Compression Texture   72
Recrystallization Textures   77
References   79

8 Substitutional Solid Solutions **80**
Phase Diagrams   80
Ternary Phase Diagrams   80
Effects of Solutes on the Eutectoid Transformation 80
Effect of Solutes on Physical Properties   81
Solid Solution Hardening   81
Carbide-Forming Tendencies   88

Solute Segregation to Grain Boundaries 89
References 89

9 Interstitial Solid Solutions 90
Atomic Diameters 90
Lattice Sites for Interstitials 90
Lattice Expansion with C, N 92
Solubility of Carbon and Nitrogen 93
Snoek Effect in bcc Metals 95
References 97

10 Diffusion 98
General 98
Mechanisms of Diffusion 99
Diffusion of Interstitials 102
References 103

11 Strain Aging 104
Yielding and Lüders Bands 104
Strain Aging 105
Dynamic Strain Aging 109
References 112

12 Austenite Transformation 113
Kinetics of Austenization 113
Pearlite Formation 113
Isothermal Transformation 120
Bainite 123
Continuous Cooling Diagrams 123
Martensite 125
Retained Austenite 126
Transformation to Martensite 128
Martensite Types 132
Special Heat Treatments 133
Miscellany 135
References 136

13 Hardenability 137
Jominy End-Quench Test 137
Hardenability Bands 140
Ideal Diameter Calculations 142

Boron   146
Miscellany   149
References   149

14 Tempering and Surface Hardening **150**
Tempering   150
Secondary Hardening   156
Temper Embrittlement   157
Carburizing   158
Carburizing Kinetics   159
Kinetics of Decarburization   162
Carboaustempering   163
Nitriding   163
Carbonitriding   164
Case Hardening Without Composition Change   165
Furnace Atmospheres   165
References   165

15 Low-Carbon Sheet Steel **167**
Sheet Steels   167
Strength   168
Grades of Low-Carbon Steel   168
Weathering Steel   174
Heating During Deformation   174
Taylor-Welded Blanks   175
Special Sheets   176
Surface Treatment   177
Special Concerns   178
References   178

16 Sheet Steel Formability **179**
Anisotropic Yielding   182
Effect of Strain Hardening on the Yield Locus   186
Deep Drawing   187
Stamping   189
Forming Limits   191
References   193

17 Alloy Steels **195**
Designation System   195
Effect of Alloying Elements   195

Applications 197
References 197

18 Other Steels **198**
Hadfield Austenitic Manganese Steel 198
Maraging Steels 198
Tool Steels 199
Heat Treatment of Tool Steels 201
Note of Interest 202
References 204

19 Stainless Steels **205**
General Corrosion Resistance 205
Ferritic Stainless Steels 205
Martensitic Stainless Steels 208
Austenitic Stainless Steels 209
Other Stainless Steels 212
Sensitization 214
Oxidation Resistance 216
References 217

20 Fracture **218**
Ductile Fracture 218
Brittle Fracture 222
Transition Temperature 226
Liquid Metal Embrittlement 227
Hydrogen Embrittlement 227
Fatigue 229
References 233

21 Cast Irons **234**
Production 234
General 234
White Irons 237
Gray Irons 237
Compact Graphite Iron 242
Ductile Cast Iron 244
Malleable Cast Iron 244
Matrices 246
Austempering of Cast Irons 250

Damping Capacity   254
References   255

22 Magnetic Behavior of Iron **256**
General   256
Ferromagnetism   257
Magnetostatic Energy   261
Magnetocrystalline Energy   262
Magnetostriction   262
Physical Units   263
The B-H Curve   264
Bloch Walls   266
Soft Versus Hard Magnetic Materials   266
Soft Magnetic Materials   266
Silicon Steel   270
Hard Magnetic Materials   272
Summary   274
References   275

23 Corrosion **276**
Corrosion Cells   276
Polarization and Passivity   279
Pourbaix Diagram   283
Types of Corrosion   284
Corrosion Control   285
Rust   287
Direct Oxidation   287
References   289

Appendix I: Physical Properties of Pure Iron **291**

**Appendix II: Approximate Hardness Conversions
and Tensile Strengths of Steels** **293**

Index   295

PREFACE

Modern civilization would not be possible without iron and steel. Steel is an essential component of all machinery used for manufacture of all our goods. The words *iron* and *steel* have come to suggest strength as evident in the following terms: *iron willed*, *iron fisted*, *iron clad*, *iron curtain*, and *pumping iron*. A *steely glance* is a stern look. A *heart of steel* implies a very hard demeanor. The Russian dictator Joseph Stalin (which means "steel" in Russian) chose that name to invoke fear in his subordinates.

This book is intended both as a resource for engineers and as an introduction to the layman to our most important metal system. The first few chapters cover the history and refining of iron and steel; the rest of the book covers physical properties and physical metallurgy.

I have drawn heavily on material from *Physical Metallurgy of Steels* by W. C. Leslie and *Steel Metallurgy for the Nonmetallurgist* by J. D. Verhoeven. However, this book includes material not covered in either of those.

Professors Robert Pehlke, Ronald Gibala, John Keough, and Paul Trojan were very helpful. Kathy Hayrynen supplied a number of micrographs.

The reader is assumed to have had a course in materials science and to be familiar with phase diagrams, Fick's laws of diffusion, and the concept of free energy.

1

GENERAL INTRODUCTION

NOMENCLATURE

The terms *iron* and *steel* are often confusing to the general public. *Iron* is an element (26 on the periodic table). The word *iron* comes from the Scandinavian word *iarn*. The chemical symbol *Fe* comes from the Latin word for iron, *ferrum*. The French word for iron is *fer*, the German word, *Eisen*. The Dutch word is *ijzeret*, and the Spanish is *hierro*.

The word *steel* is used to describe almost all alloys of iron. It is often said that steel is an alloy of iron and carbon. However, many steels contain almost no carbon. Carbon contents of some steels are as low as 0.002% by weight. The most widely used steels are low-carbon steels that have less than 0.06% carbon. Low-carbon steels are used for automobile bodies, appliances, cans, and cabinets. Higher carbon contents are used in steel with higher strengths. Tools are made from steels containing up to about 1.2% carbon.

The Sanskrit word for steel is *stakati*. The German word is *Stahl*; the Russian, *stalin*; the French, *acier*; the Spanish, *acero*; and the Dutch, *staal*. *Chalybs* is the Latin word for steel.

Wrought iron was an iron-based product with entrapped slag stringers that contained very little carbon. It is no longer produced, having been replaced by much cheaper low-carbon steel. The term *wrought iron* is still applied to garden furniture and similar products that are made today from low-carbon steel.

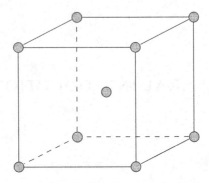

Figure 1.1. Unit cell of a body-centered cubic crystal.

Cast irons are iron-based alloys that contain 2.5 to 4% carbon and 2 to 3% silicon. In white cast iron, the carbon is present as iron carbide, whereas in gray and ductile cast irons, most of the carbon is present as graphite.

PHASES

Pure iron undergoes several phase changes. Above 1538°C, it is liquid. On cooling below 1538°C, it transforms to a body-centered cubic (bcc) structure, delta (δ)-ferrite, as shown in Figure 1.1.

On further cooling, it transforms to a denser face-centered cubic (fcc) structure, gamma (γ)-austenite, at 1400°C (Figure 1.2).

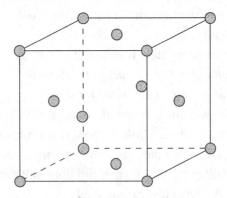

Figure 1.2. Unit cell of a face-centered cubic crystal.

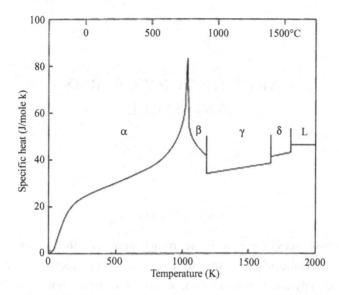

Figure 1.3. Temperature dependence of the specific heat of iron.

Below 911°C, it transforms back to bcc α-ferrite (which is the same as δ-ferrite). There is a paramagnetic-to-ferromagnetic transformation at the Curie temperature (770°C). Early researchers mistook a peak in the specific heat as a latent heat of transformation (Figure 1.3) and designated the structure between 770°C and 911°C as β-iron.

Physical properties of iron are listed in Appendix I.

PRODUCTION

The production of iron is very much greater than other metals. In the U.S. the annual tonnage of iron and steel is almost 40 times as great as that of aluminum and almost 65 times greater than the production of copper. Recycling accounts for over 80% of the steel.

2

EARLY HISTORY OF IRON
AND STEEL

NATIVE IRON

The only sources of iron available to early humans were meteoric iron and native (telluric) iron. Both were scarce. Most meteorites are non-metallic; only about 6% are iron, and these contain about 7 to 15% nickel. In 1808, William Thomson sectioned and etched a meteorite, noting the remarkable patterns. Although he published his findings in 1808, they attracted little interest. Also in 1808, an Austrian, Alois von Widmannstätten, also etched a meteorite and observed the structure that is now known by his name. In 1820, he and Carl von Schreibers published a book on meteorites, which contained a print from a heavily etched meteorite (Figure 2.1). Native iron is even scarcer, being limited to small particles in western Greenland. Archeological finds of iron with considerable amounts of nickel suggest that they were made from meteorites.

The first production of iron dates back to at least 2000 BC in India and Sri Lanka. By 1200 BC, production of iron was widespread in China and the Near East. The most common iron ores are hematite (Fe_2O_3) and magnetite (Fe_3O_4). Smelting of iron involved heating iron ore (oxides of iron) with charcoal. The reaction of iron oxide with carbon produced carbon monoxide and carbon dioxide. The air was supplied by either a natural draft or some means of blowing. Early furnaces were of various types. An open-pit furnace is shown

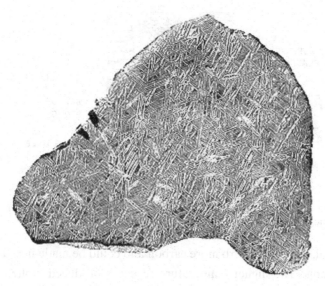

Figure 2.1. The imprint from an iron meteorite heavily etched by Widmannstätten.

in Figure 2.2. The carbon content of iron produced in pit furnaces was usually low because of the low temperatures achieved and resulted in semisolid sponge.

With shaft furnaces (Figure 2.3), the higher temperatures resulted in higher carbon contents. In the furnaces, charcoal reacted with the air to form carbon monoxide, which reduced some of the ore. The resulting carbon dioxide reacted with charcoal to form more carbon monoxide.

The product of the lower-temperature furnaces was low in carbon and much like wrought iron. It was soft and formable. If heated in

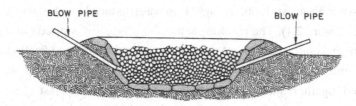

Figure 2.2. An early open-pit iron furnace. From *The Making, Shaping and Treating of Steel*, 9th ed. U.S. Steel Co. (1971).

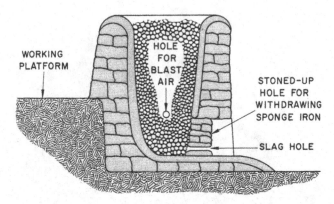

Figure 2.3. An early shaft-type iron furnace. From *The Making, Shaping and Treating of Steel*, ibid.

charcoal, it would absorb more carbon and could be made into useful steel tools. The higher-temperature furnaces produced molten iron that contained up to 4% carbon. After it solidified, it formed a brittle material that was at first discarded. Later it was learned that the carbon content could be reduced by remelting in contact with air. By 200 BC, the Chinese had started casting the high-carbon material into useful objects.

WROUGHT IRON

There is a wrought-iron pillar in Delhi, India, that dates back to at least the late fourth century. It is more than 7 m in height and has resisted corrosion over the many centuries. Wrought iron is the principal material in the Eiffel Tower, constructed in 1887.

After about AD 1300, wrought iron was produced in a Catalan furnace (Figure 2.4). The resulting semisolid product was pried out and hammered into bars. The American bloomery was a modification of this process, differing in that the charge of ore and charcoal were mixed together, and waterpower was used to create the blast.

The microstructure of wrought iron is shown in Figure 2.5, together with a typical fracture. Before the introduction of cheap

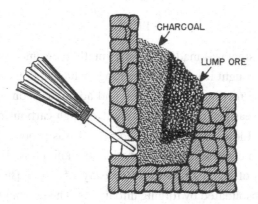

Figure 2.4. The Catalan furnace for producing wrought iron. From *The Making, Shaping and Treating of Steel*, ibid.

steel, made possible by the Bessemer process (see Chapter 3), wrought iron was used for railroad rails, locks, and about all of today's use of low-carbon steel. With the advent of cheap steel, the use of wrought iron decreased, and its last production was in the 1940s.

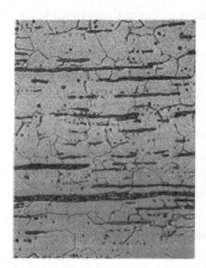

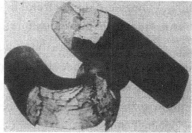

Figure 2.5. The microstructure of wrought iron (left) consists of elongated glassy silicate inclusions from slag, strung out by cold working. Fractures of wrought iron (right) occur along the silicate stringers and appear woody. From J. Aston and E. B. Stong, *Wrought Iron and Its Manufacture*, Byer Co. (1939).

STEEL

Early carbon steel was made by several methods. Some was made by carburizing wrought iron without melting by having it in contact with charcoal or by melting wrought iron and adding carbon. It was also made by decreasing the carbon content of high-carbon material by oxidation. In India and Ceylon, *wootz* steel was produced by heating iron ore in closed crucibles with charcoal and glass. This formed small buttons of high-carbon iron (typically 1.5% C). These buttons were then consolidated by forging into ingots. The technique died out about 1700 when the necessary ores containing tungsten and vanadium had been depleted. The high carbon content meant that Fe_3C was present, and this resulted in a very hard material with a characteristic surface pattern. Five-pound cakes of *wootz* steel were shipped to Persia, where they were beaten into swords. The term *Damascus steel* was given to the swords by Europeans, who first encountered their use by the Saracens in Damascus during the Crusades. The steel had 1.5% to 2.0% carbon, so it had large amounts of iron carbide, which made it very hard.

These swords were characterized by wavy patterns resembling the surface of water, the result of bands of carbide particles, which etched white, and a steel matrix, which etched black. The bands resulted from the segregation of carbide-forming elements (principally vanadium and molybdenum) during freezing. These bands attracted carbon.

An example is shown in Figure 2.6. Damascus steel sword blades were both resistant to fracture and capable of holding a very sharp edge. They were made from about AD 1100 to 1700. European attempts to reproduce these steels by forging together layers of high- and low-carbon steel were unsuccessful. It is believed that the art of producing Damascus steel involved three important factors. One was that the ore deposit contained crucial amounts of carbide-forming elements. The fact that the steels from these deposits also contained high levels of phosphorus meant that they would crack on forging

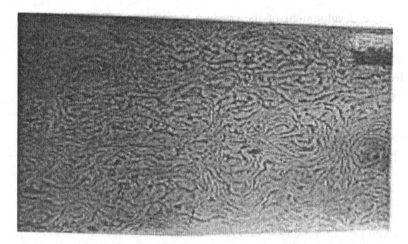

Figure 2.6. Surface of a Damascus steel sword.

unless the surface had been decarburized first. A blacksmith must also have been taught how to produce the characteristic patterns, because they would not be visible until after the decarburized surface was ground off. The secrets of making such blades were carefully guarded. Verhoeven suggested that the source of ores containing the proper impurities became unavailable, and thus the techniques stopped working.

Other ancient steel dating back to 1400 BC, comes from Africa. The people of East Africa invented a type of high-heat blast furnace that allowed them to produce carbon steel at 3275°F (1800° C) nearly 2000 years ago. This ability was not duplicated until centuries later in Europe during the Industrial Revolution.

In the fourth century BC, steel weapons were produced in Spain, The earliest production of high-carbon steel was found in Ceylon (Sri Lanka). Early steel making employed the unique use of a wind furnace, blown by the monsoon winds. This made possible the production of high-carbon steel.

Crucible steel, formed by slowly heating and cooling pure iron oxide and carbon (typically in the form of charcoal) in a crucible, was produced in central Asia by the 9th to 10th century AD.

The development of the modern blast furnace from early shaft furnaces was a gradual process. Furnaces evolved to become large and taller. The use of coke instead of charcoal was first introduced in England in 1619 but did not become common until the early 1700s. The preheated blast was first used in England in the early 1870s.

Modern steel making (Chapter 3) began with the introduction of the Bessemer process in 1856 and the open-hearth process shortly afterward.

REFERENCES

J. Aston and E. B. Stong, *Wrought Iron and Its Manufacture*, Byer Co. (1939).
The Making, Shaping and Treating of Steel, 9th ed. U.S. Steel Corp. (1971).
C. S. Smith, *A History of Metallography*, University of Chicago Press (1960).
J. D. Verhoeven, *Steel Metallurgy for the Non-Metallurgist*, ASM International (2007).

3

MODERN STEEL MAKING

BLAST FURNACE

A typical modern blast furnace is about 30 m high and 10 m in diameter (see Figure 3.1). It is lined with refractory brick. The charge consists of iron ore, coke, and limestone. Most ore is either hematite (Fe_2O_3) or magnetite (Fe_3O_4) in pieces from 0.5 to 1.5 inches in diameter. Some ores are pelletized into balls by pressing and firing very fine ore. The coke is made from coal by heating to drive off volatile oil and tar. It is screened to pieces 1 to 4 inches in diameter. Coke is much stronger than coal and contains more carbon. The limestone is in pieces from 0.5 to 1.5 inches in diameter. All of this is charged into the top of the furnace and forms layers.

Air is forced into the furnace through tuyeres near the bottom. The important reduction reactions are as follows:

$$O_2 + 2C \rightarrow 2CO$$
$$3Fe_2O_3 + CO \rightarrow CO_2 + 2Fe_3O_4 \quad \text{(at about 450°C)}$$
$$Fe_3O_4 + CO \rightarrow CO_2 + 3FeO \quad \text{(at about 600°C)}$$
$$FeO + CO \rightarrow CO_2 + Fe \quad \text{(at about 700°C)}$$

The limestone decomposes, $CaCO_3 \rightarrow CaO + CO_2$, and the CaO reacts to remove sulfur by the reaction $FeS + CaO \rightarrow CaS + FeO$. The CaS enters the slag consisting of SiO_2, Al_2O_3, MgO, and CaO, which floats on top of the molten iron. As the charge descends,

11

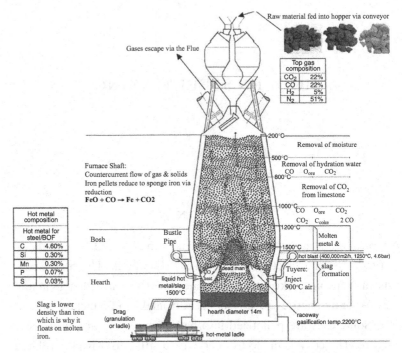

Figure 3.1. Blast furnace. Courtesy of Severstal Steel.

the gasses ascend. The composition of a typical hot metal is 4% C, 0.5% Si, 0.6% Mn, 0.04% S, 0.06% P, and 0.04% Ti.

Production of a ton of iron in a 200-ton-per-day blast furnace requires 3600 lbs of pelletized ore, 750 lbs of coke, 80 lbs of flux, 125 lbs of natural gas, 249 lbs of coal, and 3350 lbs of air.

The furnace is periodically tapped to remove the slag and recover the molten iron. Before the introduction of the Bessemer process that used molten iron, the molten metal was allowed to flow into ingot cavities carved into the sand. Because they looked like piglets suckling a sow (Figure 3.2), the product was called pig iron. These pigs were used for iron castings or to make steel in a crucible furnace with small additions of iron ore to oxidize carbon, silicon, and manganese. The crucible process was used to make steel from early in the 18th century to the mid-19th century.

From blast furnace

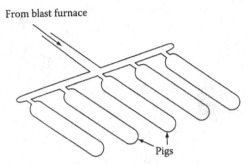

Pigs

Figure 3.2. Pig iron ingots. From W. F. Hosford, *Physical Metallurgy*, 2nd ed., CRC Press (2010).

COKE

Until about 1800, wood charcoal was used as the source of carbon. Then it was discovered that coke could be used instead. Coke is produced by baking bituminous coal in the absence of air above 2000°F to drive off particulates and volatiles such as phenol, naphthalene, ammonium sulfate, light oils and tars, and other low-molecular-weight hydrocarbons. After the coal is sized, it is heated for 14 to 36 hours. For every ton of coke, about 3 kg of SO_x, 1.5 kg of NO_x, and 3 kg of volatile hydrocarbons as well as particulates and ammonia are released and usually captured to make by-product chemicals.

BESSEMER STEEL-MAKING PROCESS

With the introduction of the Bessemer process, low-carbon steel could be cheaply mass produced, so steel quickly replaced wrought iron for most applications. The acid converter, designed by Sir Henry Bessemer in 1856, consisted of a pear-shaped vessel charged with molten pig iron. It had holes in the bottom through which air could be blown (Figure 3.3). Oxygen in the air oxidized the carbon in the iron and silicon as well. The reaction increased the temperature. Because the air blast removed too much carbon, leaving oxygen dissolved in the

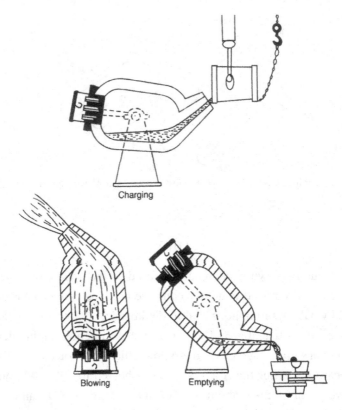

Charging

Blowing   Emptying

Figure 3.3. Bessemer converter. From C. Moore and R. I. Marshall, *Steel Making*, Fig. 2.1, p. 8, Institute of Metals (1991).

steel, spiegeleisen, which consists of iron, manganese, and some carbon, was added after the blow. The whole process took about 20 minutes to produce 20 to 30 tons of molten steel. The cheap Bessemer steel virtually replaced wrought iron for railroad rails, making the vast expansion of railroads possible in the last half of the 19th century.

Most iron ore contains phosphorus, which is not removed in the blast furnace. Phosphorus embrittles steel, so limestone was added to the converters to remove it, creating a basic process because of the calcium in the slag. The basic slag also served to remove sulfur from the steel. Bessemer steel also contains nitrogen from the air blast.

Alexander L. Holley bought the American rights to Bessemer's patent and started making steel in Troy, New York, in 1865. He improved the process with a quicker method of changing the bottom bricks.

An American, William Kelly, also did pioneering work on making steel by blowing air up through molten iron to reduce the carbon content in 1847. Kelly later asserted that English workmen at his plant had informed Bessemer of Kelly's experiments. After Bessemer had patented the process, Kelly applied for a U.S. patent, which was granted in 1857.

OPEN-HEARTH STEEL-MAKING PROCESS

The French engineer Pierre-Émile Martin took out a license in 1865 from Siemens for a regenerative furnace and applied it to making steel. The basic open-hearth process became known as the Siemens-Martin process. It allowed the rapid production of large quantities of basic steel. Most furnaces had capacities of 50 to 100 tons, but later some had a capacity of 250 or even 500 tons. The Siemens-Martin process was slower than the Bessemer process, taking several hours to reduce carbon to the desired level, and was thus easier to control. It permitted the melting and refining of large amounts of scrap steel, further lowering steel production costs. The much larger capacity compensated for the longer processing times. The basic furnace lining (principally lime and magnesia) resulted in a phosphorus content lower than that of the acid Bessemer process. The reduction took place by diffusion of oxygen ions through the slag, and thus little nitrogen was introduced because the steel was not exposed to air. The Bessemer and open-heath processes dominated steel making until the late 1960s. The last open-hearth furnace in the United States was shut down in 1992.

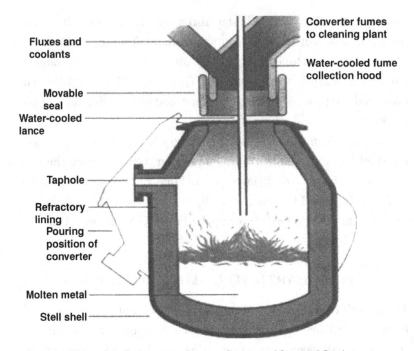

Fluxes and coolants

Converter fumes to cleaning plant

Water-cooled fume collection hood

Movable seal

Water-cooled lance

Taphole

Refractory lining

Pouring position of converter

Molten metal

Stell shell

Figure 3.4. Basic oxygen furnace. Courtesy of Severstal Steel.

BASIC OXYGEN FURNACE

The basic oxygen process was first developed in the late 1940s by a Swiss engineer, Robert Durrer. The process was first commercialized in Austria in 1952. Although the process was used in Europe and Japan, American steel companies were slow to adopt it. In 1957, McLouth Steel in Trenton, Michigan, was the first U.S. company to use the process. By 1970, 60% of the world's steel was made by this method. Figure 3.4 shows a basic oxygen furnace (BOF).

The refractory-lined furnace is first tilted 45° as shown in Figure 3.5 and charged with scrap and then hot metal (pig iron directly from the blast furnace). The ratio of hot metal to scrap (about 3–4 to 1) is critical, because no heat is supplied externally. The hot metal is pretreated to reduce the sulfur, phosphorus, and silicon content. Magnesium is added to the pig iron in the ladle to remove the sulfur. After

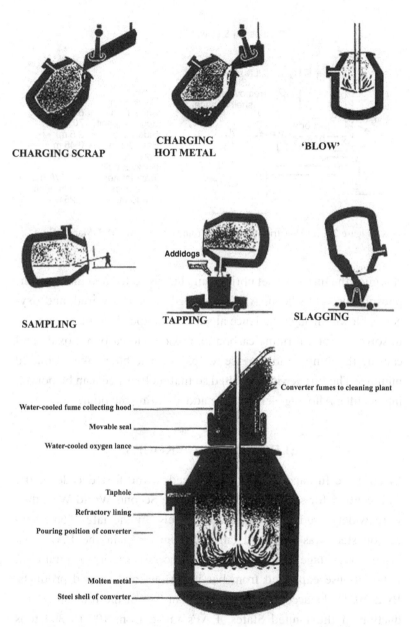

CHARGING SCRAP

CHARGING
HOT METAL

'BLOW'

SAMPLING

Addidogs

TAPPING

SLAGGING

Converter fumes to cleaning plant

Water-cooled fume collecting hood

Movable seal

Water-cooled oxygen lance

Taphole

Refractory lining

Pouring position of converter

Molten metal

Steel shell of converter

Figure 3.5. Basic oxygen furnace. From C. Moore and R. I. Marshall.

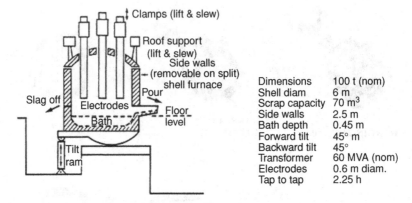

Figure 3.6. The electric arc furnace. From C. Moore and R. I. Marshall.

charging, the furnace is set upright, and lime is added to remove sulfur, phosphorus, and silicon. A water-cooled lance is lowered, and oxygen is blown through the lance at supersonic speeds to react with the dissolved carbon forming carbon monoxide and carbon dioxide and causing the temperature to rise to 1650°C. The blow takes about 20 minutes. The furnace is then tilted so that molten steel can be poured into a ladle. Alloying elements are added during the pour.

ELECTRIC ARC PROCESS

Electric arc furnaces (EAFs) were used in the first decade of the 20th century for special steels. During the Second World War, they were widely used to produce alloy steels. In the late 1960s, low-carbon steel was produced entirely from scrap in the EAF. The biggest advantage of the electric arc process is the low capital cost. Later, its use expanded from bar into sheet and shaped products. By 2010, EAFs accounted for more than half of the total steel production in the United States. EAFs range from 100 to 300 tons and require about 2 hours of processing time. Figure 3.6 illustrates an EAF.

FURNACE LININGS AND SLAGS

The Bessemer, open-hearth and BOF processes remove impurities from a mixture of pig iron and steel scrap. The impurities removed, however, depend on whether an acid (siliceous) or basic (limey) slag is used. The original Bessemer process used an acid slag and an acid furnace lining (silica), which necessitated use of a low-phosphorus ore. With an acid slag silicon, only manganese and carbon are removed by oxidation; consequently, the raw material had to be low in phosphorus, and sulfur had to be in amounts not to exceed the permissible limits in the finished steel. Later, with the introduction of a basic slag and a basic lining of magnesite or dolomite, the charge could be used with ores containing more phosphorus. In the basic processes, silicon, manganese, carbon, phosphorus, and sulfur can be removed from the charge, but normally the raw material contains low silicon and high phosphorus contents. To remove the phosphorus, the bath of metal must be oxidized to a greater extent than in the corresponding acid process, and the final quality of the steel depends largely on the degree of this oxidation. This must be done before deoxidizers, ferro-manganese, ferro-silicon, and aluminum are added. Aluminum removes the soluble oxygen: $2Al + 3FeO$ (soluble) $\rightarrow 3Fe + Al_2O_3$ (solid).

In acid processes, deoxidation can take place in the furnaces, leaving a reasonable time for the inclusions to rise into the slag and thus be removed before casting. In basic furnaces, deoxidation is rarely carried out in the presence of the slag; otherwise phosphorus would return to the metal. Deoxidation of the metal takes place in the ladle, leaving only a short time for the deoxidation products to be removed. For these reasons, acid steel was considered better than basic steel for certain purposes, such as large forging ingots and ball-bearing steel. The introduction of vacuum degassing hastened the decline of the acid processes. Lime dolomite and magnesia are basic refractories, and graphite, chromite, and bone ash are neutral refractories. Silica can tolerate temperatures as high as 1750°C, alumina up to 2000°C, and magnesia up to 2200°C.

One measure of the basicity of a slag or furnace lining is the V-ratio: $(\%CaO - 1.18\%P_2O5)/\%SiO_2$.

CASTING

As soon as the steel is produced, it must be cast. Through the 1950s, steel was cast into molds. Continuous casting of steel was not common until the 1960s. If no treatment was done, dissolved carbon and oxygen would react violently as the metal froze to produce carbon monoxide, $C + O \rightarrow CO$. The evolving carbon monoxide would cause small droplets of iron to be ejected, and these would burn, creating a display. This was called *rimming* and the ingot product was called *rimming steel*. Alternatively, aluminum might be added in the ladle to react with the dissolved oxygen to form Al_2O_3. The freezing would then be quiet because the steel was said to have been *killed*, or to have an oxygen level low enough that no CO would evolve during solidification.

Today, approximately 96% of steel is continuously cast with only 4% being cast into ingots. The steel industry realized that production and quality could be improved by continuous casting. The first continuous casters were vertical, but to reduce the height required, casting into curves became standard (Figure 3.7). Molten metal from steel making is killed in the ladle with aluminum to remove the oxygen. Control over the process is critical: the rate of pouring must match the rate at which the ingot is withdrawn, and this must be controlled by how fast the material freezes. There must also be a constant supply of hot metal. If the process is stopped, a new tundish is required because metal will have frozen in the old one. Start-up requires the use of a starter bar in the bottom of the mold.

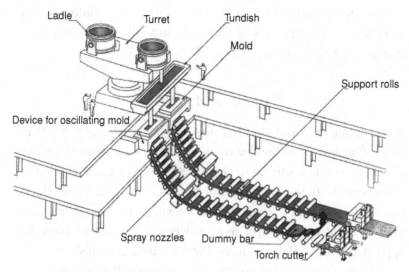

Figure 3.7. Typical continuous casting. Courtesy of Severstal Steel.

As the billet leaves the mold, it is sprayed with water to solidify it before it is cut into slabs, billets, or blooms by a moving oxyacetylene torch.

HOT ROLLING

Continuously cast billets are generally cut into lengths that are then hot rolled, either into final shapes or into plate. Shapes such as railroad rails, I beams, and bars are rolled through reversing mills, with each pass progressively shaping the product. Flat products are rolled continuously through a series of rolls, which gradually reduce the thickness.

Hot rolling is defined as rolling above the recrystallization temperature (see Chapter 6). For steel, it is usually started at 1100°C but finishes at a much lower temperature. After hot rolling, the steel is pickled to remove oxide scale.

For some products, such as I beams, railroad rails, reinforcing rod, and plates, the last processing step is the hot rolling. However, most steel is cold rolled into sheet.

COLD ROLLING

It is common practice to hot roll steel to a thickness of about 0.25 in. Hot rolling has the advantage of lower rolling forces, but for thinner plates and sheets, frictional forces become important, and lubrication is not possible. Below thicknesses of about 0.25 in, further reduction is usually done cold to the final desired thickness. Thickness reductions of 85% result in the gauges most widely used for automobiles and appliances. Cold rolling produces a very good surface finish.

Almost all cold-rolled steel is given a recrystallization anneal to make it more formable. This must be done in a controlled atmosphere to protect the surface from oxidation. Usually the sheet is then either temper rolled or roller leveled to remove the yield point. This reduces the thickness less than 0.002 in. Finally, most sheets are given a surface treatment before shipping.

A large fraction of cold-rolled steel is galvanized (plated with zinc) for corrosion resistance. Galvanizing may be done by depositing zinc from a molten bath (hot dipping), as illustrated in Figure 3.8. Alternatively, much galvanizing is done by electroplating (Figure 3.9).

RECYCLING

In the United States, more steel is recycled each year than aluminum, copper, plastics, paper, and glass combined. Of the total U.S. steel production, approximately 75% is from recycled steel. The BOF uses from 25% to 35% scrap, and the EAF uses 100%. Approximately 17% of the recycled material is offal (material returned from manufacturing), with the other 83% being from used consumer products.

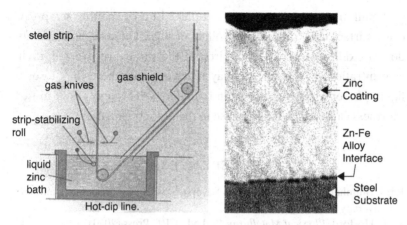

Figure 3.8. Hot dipping line (left) and microstructure of the surface (right). Courtesy of Severstal Steel.

Steel and iron can easily be separated magnetically from other scrap. The scrap is either remelted in an EAF or added to pig iron in a BOF. All grades of steel can be recycled because most alloying elements are oxidized during processing. Tin and copper are the exceptions, and there is concern in the steel industry about the gradual buildup of these *tramp* elements in steel. This problem arises from recycling of automobile bodies from which wiring has not been removed.

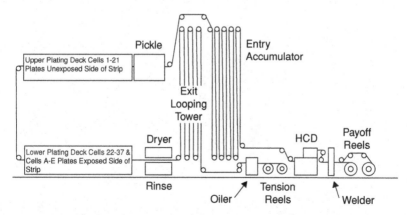

Figure 3.9. Electrogalvanizing line. Courtesy of Severstal Steel.

Small amounts of copper or tin picked up by melting scrap can cause surface cracks during hot rolling of steel. These tramp elements do not oxidize easily, so as the iron at the surface is oxidized, their concentrations at the surface may reach a point where they form a liquid phase, which wet the grain boundaries. This may result in hot shortness caused by a liquid phase in the grain boundaries.

REFERENCES

Beth Blumhardt, presentation for Severstal Steel, departmental seminar, University of Michigan (2011).

W. F. Hosford, *Physical Metallurgy*, 2nd ed., CRC Press (2010).

C. Moore and R. I. Marshall, *Steel Making*, Institute of Metals (1991).

4

CONSTITUTION OF
CARBON STEELS

MICROSTRUCTURES OF CARBON STEELS

Steels are iron-based alloys. The most common are carbon steels, which may contain up to 1.5% carbon. Cast irons typically contain between 2.5 and 4% carbon. Figure 4.1 is the phase diagram showing the metastable equilibrium between iron and iron carbide. Below 912°C, pure iron has a body-centered cubic (bcc) crystal structure and is called *ferrite*, which is designated by the symbol α. Between 912 and 1400°C, the crystal structure is face-centered cubic (fcc). This phase, called *austenite*, is designated by the symbol γ. Between 1400°C and the melting point, iron is again bcc. This phase is called δ-ferrite, but it is really no different from α-ferrite. The maximum solubility of carbon in α (bcc iron) iron is 0.02% C and in γ (fcc iron) is about 2%. Iron carbide, Fe_3C, is called *cementite* and has a composition of 6.67% C. The structure developed by the eutectoid reaction, $\gamma \rightarrow \alpha + Fe_3C$ at 727°C, consists of alternating platelets of ferrite and carbide (Figure 4.2) and is called *pearlite*.

Steels containing less than 0.77% C are called *hypoeutectoid*, and those with more than 0.77% C are called *hypereutectoid*. The microstructures of medium-carbon steels (0.2 to 0.7% C) depend on how rapidly they are cooled from the austenitic temperature. If the cooling is very slow (furnace cooling), the *proeutectoid* ferrite will form in the austenite grain boundaries, surrounding regions

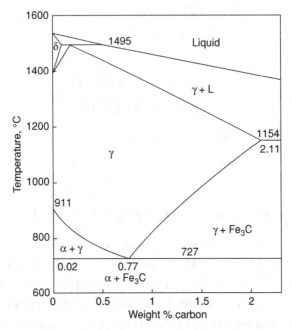

Figure 4.1. The iron-iron carbide phase diagram. Data from J. Chipman, *The Making, Shaping and Treating of Steel*, 9th ed., U.S. Steel Corp. (1971).

of austenite that subsequently transforms to pearlite, as shown in Figure 4.3.

With somewhat more rapid cooling (air cooling), there is not enough time to allow all of the carbon to diffuse from the austenite grain boundaries into the center of the austenite grains. The proeutectoid ferrite will form as plates (which appear as needles in a two-dimensional microstructure) that penetrate into the centers of the old austenite grains. This decreases the distance the carbon must diffuse. The resulting microstructure is often called a Widmanstätten structure (Figure 4.4).

For a hypoeutectoid steel, the term *full anneal* is used to describe heating 25 to 30° into the austenite field followed by a furnace cool. This produces coarse pearlite. Heating 50 to 60° into the austenite range followed by air cooling is called *normalizing*. It produces a much

Figure 4.2. Pearlite consisting of alternating platelets of ferrite and iron carbide. From *The Making, Shaping and Treating of Steel*, 9th ed., ibid.

finer pearlite. *Spheroidizing* is a heat treatment used on medium-carbon steels to improve cold formability. This can be accomplished by heating just below the eutectoid temperature for 1 or 2 hours. At this temperature, surface tension causes the carbides to spheroidize. Some companies use a more complicated process that involves first heating above the eutectoid temperature for 2 hours and then slowly cooling below it and holding 8 hours or more. The reason this process is used is unclear. Spheroidizing is much faster without heating into the austenitic region.[*]

All carbon steels contain manganese and silicon in addition to carbon. Silicon and manganese are present as impurities. Manganese reacts with sulfur, which is always present as an impurity, to form MnS. If manganese were not present, the sulfur would form FeS. Iron

[*] J. O'Brien and W. Hosford, "Spheroidization Cycles for Medium Carbon Steels," *Met. and Mat. Trans.*, v. 33 (2002), pp. 175–177.

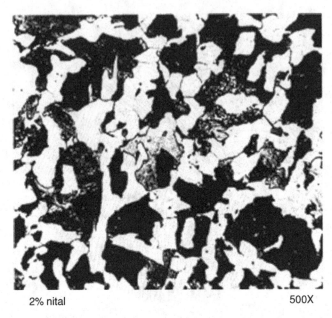

2% nital   500X

Figure 4.3. Hypoeutectoid 1045 steel normalized by air cooling. The dark regions are pearlite, and the light regions are ferrite. Note the proeutectoid ferrite formed in the prior austenite grain boundaries. From *ASM Metals Handbook*, 8th ed. v. 7, ASM (1972).

sulfide wets the austenite grain boundaries and is molten at the hot working temperatures. This results in the steel being hot short (having no ductility at high temperatures).

When molten steel first freezes, the manganese and other alloying elements tend to lie in the interdendritic regions. Carbon is more strongly attracted to manganese than to iron. If the steel is hot rolled into plate or tube, the initial regions of interdendritic segregation become aligned with the rolling direction. The proeutectoid ferrite will form in the carbon-lean regions and the pearlite in the carbon-rich regions. The resulting microstructure will have a very directional microstructure, called a *banded* microstructure (Figure 4.5).

The directionality of the microstructure would suggest that the ductility transverse to the rolling direction would be worse than that parallel to the rolling direction. It is. However, the principal reason

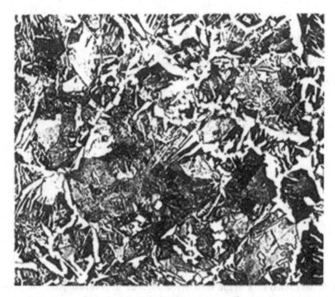

Figure 4.4. Widmanstätten structure formed in a hypoeutectoid steel during a rapid air cool. From *ASM Metals Handbook*, ibid.

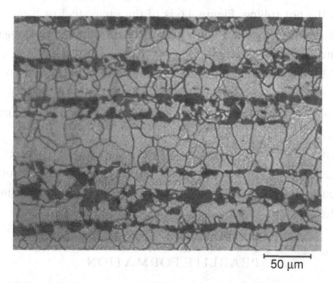

Figure 4.5. Banded microstructure of a 1018 steel slowly cooled from 900°C. The white areas are ferrite sand, and the dark areas are pearlite. From J. D. Verhoeven, *Steel Metallurgy for the Nonmetallurgist*, ASM (2007).

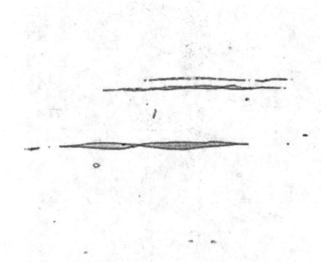

Figure 4.6. Magnified view of the structure in Figure 5 as polished. Note the elongated manganese sulfide stringers. From *ASM Metals Handbook*, ibid.

is not the banding, but the elongation of MnS inclusions that occurs during the hot rolling. Figure 4.6 is of the same steel as Figure 4.5, unetched and viewed at a much higher magnification. The elongated MnS inclusions are apparent.

Additional manganese is often added to promote lower ductile-brittle transition temperatures (see Chapter 19). Silicon is present as an impurity in amounts up to 0.25%. Other common impurities include S and P in amounts up to 0.05%.

Figure 4.7 shows the microstructure of a slowly cooled hypereutectoid steel. Proeutectoid cementite forms at the austenite grain boundaries on cooling above 727°C. The rest of the structure transforms to pearlite.

PEARLITE FORMATION

Figure 4.8 shows the eutectoid reaction in Fe-C alloys. On cooling through the eutectoid temperature, pearlite is formed by the reaction

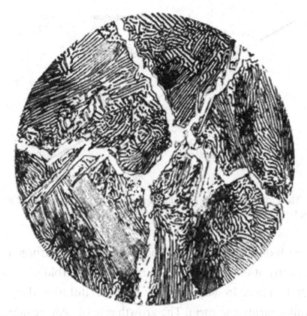

Figure 4.7. Hypereutectoid steel. Note the proeutectoid cementite formed in the prior austenite grain boundaries. From *The Making, Shaping and Treating of Steel*, ibid.

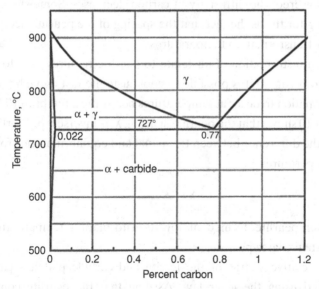

Figure 4.8. Eutectoid transformation in the iron-carbon system. From W. F. Hosford, *Physical Metalurgy*, 2nd ed., CRC (2010).

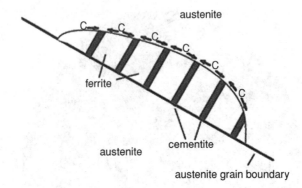

Figure 4.9. The growth of pearlite into austenite requires carbon to diffuse away from the growing ferrite toward the growing carbide. From W. F. Hosford, ibid.

$\gamma \to \alpha +$ carbide. At the austenite-pearlite interface, carbon must diffuse from in front of the ferrite to in front of the carbide, as shown in Figure 4.9. In a pure iron carbon alloy, it is this diffusion that controls how rapidly pearlite can form. The growth rate of pearlite depends on the temperature at which the austenite transforms. As the temperature is lowered, the diffusivity of carbon decreases. Somewhat compensating for this is the fact that the spacing of the pearlite decreases at lower transformation temperatures.

The transformation of austenite to pearlite requires time for diffusion to occur, so it is possible to cool austenite fast enough for the transformation to occur at temperatures below the equilibrium 727°C. Figure 4.10 shows that the pearlite spacing, λ, is inversely proportional to ΔT, the difference between the actual and equilibrium transformation temperatures:

$$\lambda = K\Delta T. \tag{4.1}$$

The finer pearlite formed at low transformation temperatures is harder than coarse pearlite.

The relative widths of the ferrite and carbide platelets can be calculated using the lever law. Assume that the pearlite contains

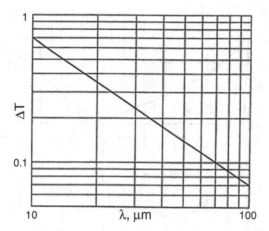

Figure 4.10. The lamellae spacing in pearlite is finer the lower the transformation temperature. Data from D. D. Pearson and J. D. Verhoeven, *Met. Trans. A*, v. 15A (1984). From W. F. Hosford, ibid.

0.77% C, ferrite 0.02%, and carbide 6.67%. The weight fraction ferrite, $f_\alpha = (6.67 - 0.77)/(6.67 - 0.02) = 0.889$, so the fraction carbide $= 0.111$. The densities of ferrite and carbide are almost equal, thus volume fractions are nearly identical to the weight fractions. The ratio of the thicknesses of the ferrite and carbide lamellae is therefore $0.889/0.111 = 7.99$.

When a hypoeutectoid steel undergoes the transformation at a temperature below the equilibrium transformation temperature, carbide cannot precipitate until the austenite is saturated with respect to carbide. This is illustrated in Figure 4.11. Consider a steel containing 0.40% C undergoing transformation at 650°C. An extrapolation of the line representing the solubility of carbon in austenite (the Hultgren extrapolation) indicates that carbon cannot precipitate from the austenite. Initially, only ferrite can form. However, the precipitation of ferrite leaves the remaining austenite enriched in carbon. Carbide will form once the carbon content reaches 0.62%, so the pearlite that forms will contain 0.62% carbon. The ratio of the thickness of the

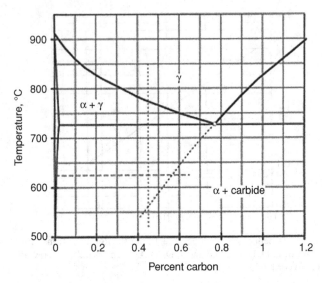

Figure 4.11. The Hultgren extrapolation of the solubility of austenite for carbon. At 625°C, carbide cannot form until the carbon content of the austenite reaches 0.57% C.

ferrite and carbide platelets in the pearlite will be greater than the value of 7.99 calculated for transformation at 727°C.

REFERENCES

ASM Metals Handbook, 8th ed. v. 7, ASM (1972).
W. F. Hosford, *Physical Metallurgy*, 2nd ed., CRC Press (2010).
A. Hultgren, *Discussion in Hardenability of Alloy Steels*, ASM (1938).
The Making, Shaping and Treating of Steel, 9th ed., U.S. Steel Corp. (1971).
J. D. Verhoeven, *Steel Metallurgy for the Nonmetallurgist*, ASM (2007).

5

PLASTIC STRENGTH

DISLOCATION DENSITY

Strain hardening is a result of the increased dislocation density with straining. Figure 5.1 shows how dislocation density increases with strain. The increased dislocation density in turn causes the flow stress to increase, as shown in Figure 5.2. This increase of flow stress with the square root of dislocation density, ρ, can be expressed as

$$\sigma = \sigma_0 + \alpha Gb\sqrt{\rho}, \tag{5.1}$$

where σ is the flow stress, G is the shear modulus, b is the Burgers vector, and σ_0 and α are constants. The strain rate in a crystal is given by

$$\dot{\varepsilon} = 0.5b\rho\bar{v}, \tag{5.2}$$

where $\bar{v}$ is the average dislocation velocity. In body-centered cubic (bcc) metals, edge dislocation moves five times as fast as screws.

STRAIN HARDENING

Often the strain hardening can be approximated by a power law:

$$\sigma = K\varepsilon^n, \tag{5.3}$$

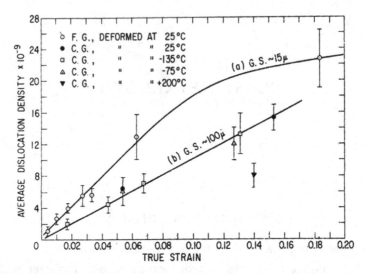

Figure 5.1. Dislocation density increases linearly with plastic strain. From A. S. Keh and S. Weissmann, *Electron Microscopy and Strength of Crystals*, Interscience (1963).

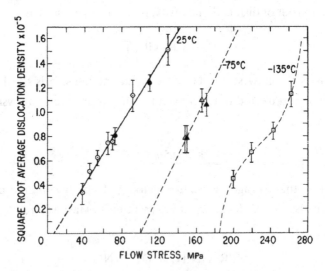

Figure 5.2. The flow stress is proportional to the square root of the dislocation density. From A. S. Keh and S. Weissmann, ibid.

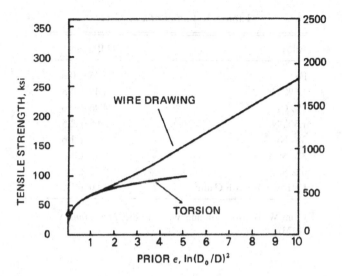

Figure 5.3. Strain hardening of Fe-0.15% Ti. From W. C. Leslie, *The Physical Metallurgy of Steels*, Hemisphere (1981).

where σ is the true flow stress and ε is the true strain. Morrison suggested that the strain-hardening exponent, n, could be approximated by

$$n = 5/(10 + d^{-1/2}). \qquad (5.4)$$

For drawn wires, however, work hardening is linear, as shown in Figure 5.3.

GRAIN SIZE

The effect of grain size on flow stress, σ, is well described by the Hall-Petch relationship,

$$\sigma = \sigma_0 + kd^{-1/2}, \qquad (5.5)$$

where d is the grain diameter and σ_0 and k are constants. Table 5.1 lists the dependence of the constant, k, on solutes.

Table 5.1. *Values of k at 22°C*

Alloy	k (MPa/mm$^{-1/2}$)
Fe	5.45 ± 0.8
1.5% Cr	5.45 ± 0.8
3% Cr	4.96 ± 0.6
1.5% Ni	9.65 ± 2.5
3% Ni	15.6 ± 1.6
3% Si	9.45 ± 1.5
6% Si	9.52 ± 1.5
Ferrovac E (with C and N)	18.0 ± 1.2

From W. B. Morrison and W. C. Leslie, *Met. Trans.* v. 4 (1973).

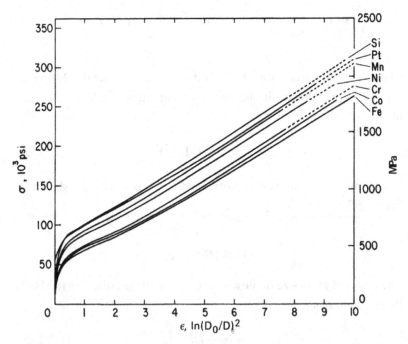

Figure 5.4. The effect of solutes on the tensile strength of drawn wires is independent of the prior strain. From G. Langford, P. K. Nagata, R. J. Sober, and W. C. Leslie, Plastic Flow in Binary Substitutional Alloys of BCC Iron–Effecting of Wire Drawing and Alloy Content on Work Hardening and Ductility. *Met. Trans.*, v. 3 (1972).

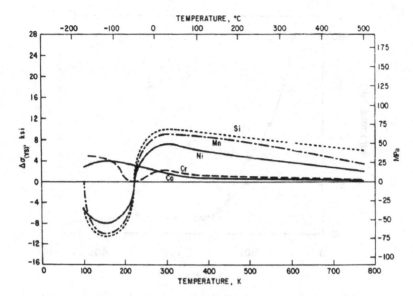

Figure 5.5. Solid solution softening below about 200 K and solid solution strengthening at higher temperatures. From W. C. Leslie, *Met. Trans.* v. 3 (1972).

SOLUTE EFFECTS

Figure 5.4 shows that that the effect of substitutional solutes on the tensile strength of drawn wire is independent of the strain hardening. At low temperatures, solutes may actually cause solid solution softening, as shown in Figure 5.5.

TEMPERATURE DEPENDENCE

The flow stress of iron is temperature dependent below room temperature, as shown in Figure 5.6.

Pickering[*] has suggested that the effects of solutes and grain size are additive. He proposed

$$YS = 15.4[3.5 + 21(\%Mn) + 5.4(\%Si)$$
$$+ 23\sqrt{(\%Ni)} + 1.13d^{-1/2}] \qquad (5.6)$$

[*] F. B. Pickering, *Towards Improved Toughness and Ductility*, Climax Molybdenum (1971).

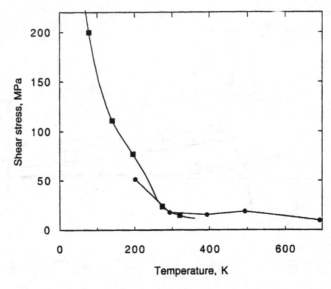

Figure 5.6. The temperature dependence of yield strength and flow stress of iron. From William F. Hosford "Finite, Deformating and Fractured" in *Encyclopedia of Materials Science and Technology*, Elsevier Science Ltd. (2001).

and

$$TS = 15.4[19.1 + 1.8(\%Mn) + 5.4(\%Si) + 23\sqrt{(\%Ni)}$$
$$+ 0.25(\% \text{ pearlite}) + 0.5d^{-1/2}], \tag{5.7}$$

where the yield strength (YS) and tensile strength (TS) are in MPa and d is in mm.

HARDNESS

The tensile strength and hardness correlate. The tensile strength in MPa is approximately three times the Vickers hardness (see Appendix II).

STRAIN-RATE DEPENDENCE OF FLOW STRESS

The average strain rate during most tensile tests is in the range of 10^{-3} to 10^{-2}/s. If it takes 5 minutes during the tensile test to reach a strain

Table 5.2. *Typical values of the strain-rate exponent, m,
at room temperature*

Material	m
Low-carbon steels	0.010 to 0.015
High-strength low-alloy steels	0.005 to 0.010
Austenitic stainless steels	-0.005 to $+0.005$
Ferritic stainless steels	0.010 to 0.015

of 0.3, the average strain rate is $\dot{\varepsilon} = 0.3/(5 \times 60) = 10^{-3}$/s. A strain
rate of $\dot{\varepsilon} = 10^{-2}$/s; a strain of 0.3 will occur in 30 seconds. For many
materials, the effect of the strain rate, $\dot{\varepsilon}$, on the flow stress, σ, at a fixed
strain and temperature can be described by a power-law expression,

$$\sigma = C\dot{\varepsilon}^m, \tag{5.8}$$

where the exponent, m, is called the *strain-rate sensitivity*. The relative
levels of stress at two strain rates (measured at the same total strain)
is given by

$$\sigma_2/\sigma_1 = (\dot{\varepsilon}_2/\dot{\varepsilon}_1)^m, \tag{5.9}$$

or $\ln(\sigma_2/\sigma_1) = m\ln(\dot{\varepsilon}_2/\dot{\varepsilon}_1)$. If σ_2 is not much greater than σ_1,

$$\ln(\sigma_2/\sigma_1) \approx \Delta\sigma/\sigma, \tag{5.10}$$

where $\Delta\sigma = (\sigma_2 - \sigma)$ and $\sigma = (\sigma_2 \, \sigma)$.
    Equation 5.9 can be simplified to

$$\Delta\sigma/\sigma \approx m\ln(\dot{\varepsilon}_2/\dot{\varepsilon}_1) = 2.3m\log(\dot{\varepsilon}_2/\dot{\varepsilon}_1). \tag{5.11}$$

Ratios of σ_2/σ_1 calculated from Equation 5.10 are shown for various
levels of $\dot{\varepsilon}_2/\dot{\varepsilon}_1$ and m in Figure 5.7.
    At room temperature, the values of m for most engineering met-
als are between -0.005 and $+0.015$; m-values for steels are shown in
Table 5.2.
    Consider the effect of a 10-fold increase in strain rate, ($\dot{\varepsilon}_2/\dot{\varepsilon}_1 =$
10) with $m = 0.01$. Equation 5.11 predicts that the level of the stress

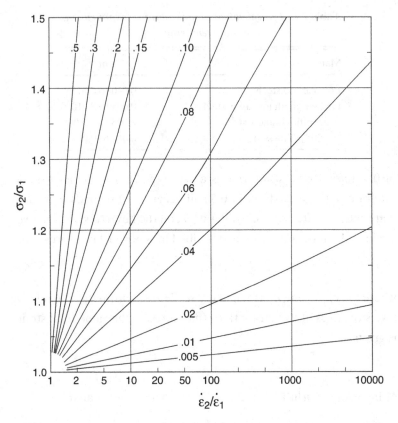

Figure 5.7. Influence of strain rate on flow stress for various m-values. From W. F. Hosford and R. M. Caddell, *Metal Forming: Mechanics and Metallurgy*, 4th ed., Cambridge University Press (2011).

increases by only $\Delta\sigma/\sigma = 2.3(0.01)(1) = 2.3\%$. This increase is typical of room temperature tensile testing. It is so small that the effect of strain rate is often ignored. Figure 5.7 shows how the relative flow stress depends on strain rate for several levels of m. The increase of flow stress, $\Delta\sigma/\sigma$, is small unless either m or $(\dot{\varepsilon}_2/\dot{\varepsilon}_1)$ is high.

Figure 5.8 illustrates two ways of determining the value of m. One method is to run two continuous tensile tests at different strain rates and compare the levels of stress at some fixed strain. The other way is to change the strain rate suddenly during a test and compare the

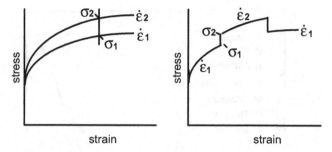

Figure 5.8. Two methods of determining the strain-rate sensitivity. Either continuous stress-strain curves at different strain rates can be compared at the same strain (left) or sudden changes of strain rate can be made and the stress levels just before and just after the change compared (right). In both cases, Equation 5.9 can be used to find m. In rate-change tests, $(\dot{\varepsilon}_2/\dot{\varepsilon}_1)$ is typically 10 or 100. From W. F. Hosford, *Mechanical Behavior of Materials*, 2nd ed., Cambridge University Press (2010).

levels of stress immediately before and after the change. The latter method is easier and therefore more common. The two methods may give somewhat different values for m. In both cases, Equation 5.9 can be used to find m. In rate-change tests, $(\dot{\varepsilon}_2/\dot{\varepsilon}_1)$ is typically 10 or 100.

For most metals, the value of the rate sensitivity, m, is low, near room temperature, but increases with temperature. The increase of m with temperature is quite rapid above half of the melting point $(T > T_m/2)$ on an absolute temperature scale. In some cases, m may be 0.5 or higher. Figure 5.9 shows the temperature dependence of m for several metals. For some alloys, there is a minimum between $0.2T_m$ and $0.3T_m$. For some metals, the rate sensitivity is slightly negative in the temperature range at about one third of the melting point. For iron, this is about 330°C.

For bcc metals, including steels, a better description of the strain-rate dependence of flow stress is given by

$$\sigma = C + m' \ln(\dot{\varepsilon}), \qquad (5.12)$$

in which case a change of strain rate from $\dot{\varepsilon}_1$ to $\dot{\varepsilon}_2$ raises the flow stress by $\Delta\sigma = m\sigma \ln(\dot{\varepsilon}_2/\dot{\varepsilon}_1) = m' \ln(\dot{\varepsilon}_2/\dot{\varepsilon}_1)$, so

$$\Delta\sigma = m' \ln(\dot{\varepsilon}_2/\dot{\varepsilon}_1), \qquad (5.13)$$

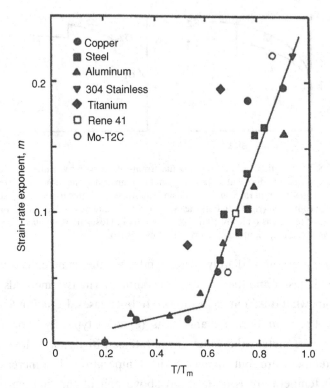

Figure 5.9. Variation of the strain-rate sensitivity, m, with temperature for several metals. Above about half of the melting point, m rises rapidly with temperature. From W. F. Hosford and R. M. Caddell, ibid.

which is independent of the stress level. Figure 5.10 shows that the stress-strain curves of iron are raised by a constant level, $\Delta\sigma$, that is independent of the stress level.

Figure 5.11 shows that the conventional m-value is dependent on the stress level. Comparing Equations 5.7 and 5.11, $\Delta\sigma = m\sigma \ln(\dot{\varepsilon}_2/\dot{\varepsilon}_1) = m' \ln(\dot{\varepsilon}_2/\dot{\varepsilon}_1)$, which indicates that $m' = m\sigma$. If a smooth curve is drawn though the data in Figure 5.9, it is apparent that $m' = 10$ MPa.

Increased strain rates lower the strain-hardening exponent n.

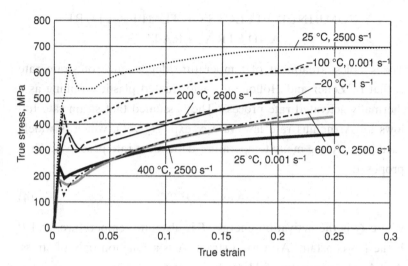

Figure 5.10. Stress-strain curves for iron at 25°C. Note that the difference in the level of the curves is independent of the stress level. From G. T. Gray in *ASM Metals Handbook*, v. 8 (2000).

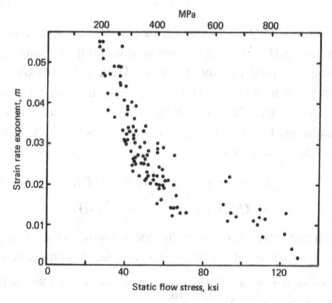

Figure 5.11. Decrease of conventional strain-rate exponent, *m*, with stress. Data from A. Saxena and D. D. Chatfield, SAE paper 760209, 1976.

COMBINED EFFECTS OF TEMPERATURE
AND STRAIN RATE

The simplest treatment of temperature dependence of strain rate is that of Zener and Hollomon,[*] who treated plastic straining as a thermally activated rate process. They assumed that strain rate follows an Arrhenius rate law, rate $\propto \exp(-Q/RT)$, which is widely used in analyzing many temperature-dependent rate processes. They proposed

$$\dot{\varepsilon} = A \exp(-Q/RT), \qquad (5.14)$$

where Q is the activation energy, T is absolute temperature, and R is the gas constant. At a fixed strain, A is a function only of stress, $A = A(\sigma)$, so Equation 5.11 can be written

$$A(\sigma) = \dot{\varepsilon} \exp(+Q/RT) \qquad (5.15)$$

$$\text{or } A(\sigma) = Z, \qquad (5.16)$$

where $Z = \dot{\varepsilon}\exp(+Q/RT)$ is called the Zener-Hollomon parameter.[*]

The Zener-Hollomon development is useful if the temperature and strain-rate ranges are not too large. Dorn and coworkers measured the activation energy, Q, for pure aluminum over a very large temperature range. They did this by observing the change of creep rate, under fixed load when the temperature was suddenly changed. Because the stress is constant,

$$\dot{\varepsilon}_2/\dot{\varepsilon}_1 = \exp[(-Q/R)(1/T_2 - 1/T_1)] \qquad (5.17)$$

$$\text{or } Q = R\ln(\dot{\varepsilon}_2/\dot{\varepsilon}_1)/(1/T_1 - 1/T_2). \qquad (5.18)$$

Dorn and coworkers[*] found that for aluminum, Q was independent of temperature above 500 K, but very temperature dependent

---

[*] C. Zener and H. Hollomon, "Effect of Strain Rate upon the Plastic Flow of Steel," *J. Appl. Phys.* v. 15 (1944), pp. 1073–1084.

[*] O. D. Sherby, J. L. Lytton, and J. E. Dorn, "The Activation Energies for Creep of Single Aluminum Crystals Favorably Oriented for (III)[101] Slip," *AIME Trans.*, v. 212 (1958), pp. 220–225.

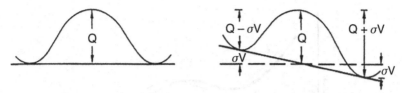

Figure 5.12. Schematic illustration of the skewing of an activation barrier by an applied stress. From W. F. Hosford and R. M. Caddell, ibid.

at lower temperatures. This observation was later explained by Z. S. Basinski and others. The basic argument, which applies to all metals, is that the stress helps thermal fluctuations overcome activation barriers. If there is no stress, the activation barrier has a height of Q, with random fluctuations. The rate of overcoming the barrier is proportional to $\exp(-Q/RT)$. An applied stress skews the barrier so that the effective height of the barrier is reduced to $Q - \sigma v$, where v is a parameter with the units of volume. Now the rate of overcoming the barrier is proportional to $\exp[-(Q - \sigma v)/RT]$. The finding of lower activation energies at lower temperatures was explained by the fact that in the experiments, greater stresses were applied at lower temperatures to achieve measurable creep rates.

In Figure 5.12, the rate of the overcoming the barrier from left to right is proportional to $\exp[-(Q - \sigma v)/RT]$, whereas the rate from right to left is proportional to $\exp[-(Q + \sigma v)/RT]$. Thus the net reaction rate is

$$C\{\exp[-(Q-\sigma v)/RT] - \exp[-(Q+\sigma v)/RT]\}$$
$$= C\exp(-Q/RT)\{\exp[(\sigma v)/RT] - \exp[-(\sigma v)/RT]\}. \quad (5.19)$$

This simplifies to $\dot{\varepsilon} = 2C\exp(-Q/RT)\sinh[(\sigma v)/RT]$.

Equation 5.8 has been modified, based on some theoretical arguments, to provide a better fit with experimental data,

$$\dot{\varepsilon} = A\exp(-Q/RT)[\sinh(\alpha\sigma)]^{1/m}, \quad (5.20)$$

where α is an empirical constant and the exponent $1/m$ is consistent with Equation 5.11.

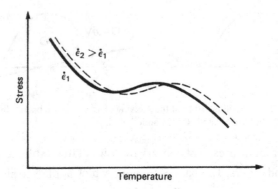

Figure 5.13. Increasing a strain rate has the same effect as lowering the temperature. From W. F. Hosford and R. M. Caddell, ibid.

Figure 5.13 shows that Equation 5.20 can be used to correlate the combined effects of temperature, stress, and strain rate over an extremely large range of strain rates.

If $\alpha\sigma \ll 1$, $\sinh(\alpha\sigma) \approx \alpha\sigma$, Equation 5.20 simplifies to

$$\dot{\varepsilon} = A \exp(+Q/RT)(\alpha\sigma)^{1/m}, \tag{5.21}$$

$$\sigma = \dot{\varepsilon}^m A' \exp(-mQ/RT), \tag{5.22}$$

and

$$\sigma = A'Z^m, \tag{5.23}$$

where $A' = (\alpha A m)^{-1}$. Equation 5.23 is consistent with the Zener-Hollomon development. At low temperatures and high stresses, $(\alpha\sigma) \gg 1$, so $\sinh(\alpha\sigma) \to \exp(\alpha\sigma)/2$ and Equation 5.20 reduces to

$$\dot{\varepsilon} = C \exp(\alpha'\sigma - Q/RT). \tag{5.24}$$

Under these conditions, both C and α' depend on strain and temperature. For a constant temperature and strain,

$$\sigma = C + m' \ln \dot{\varepsilon}. \tag{5.25}$$

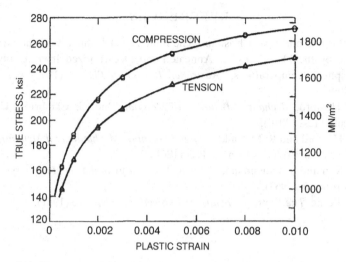

Figure 5.14. The strength differential in an as-quenched 4340 steel. From G. C. Rauch and W. C. Leslie, "The Extent and Nature of the Strength-Differential Effect in Steels," *Met. Trans.*, v. 3 (1972).

Note that this equation is consistent with Equation 5.11, but not with the power-law rate expression, Equation 5.7.

Dynamic strain aging is covered in Chapter 11.

SUPERPLASTICITY

Superplasticity is characterized by extremely high elongation and by very low flow stresses. The very low flow stress can be achieved in iron by cycling through the $\alpha \rightarrow \gamma$ transition. Tensile elongations of 300% have been achieved.

STRENGTH DIFFERENTIAL EFFECT

It has been found that as-quenched steels have higher true stress-strain curves in compression than in tension. Figure 5.14 shows that for an as-quenched 4340 steel, the compression stress-strain curve is about 8% higher than the tension. The reason for this effect is not well understood.

REFERENCES

Z. S. Basinski and J. W. Christian, "The Influence of Temperature and Strain Rate on the Flow Stress of Annealed and Decarburized Iron at Subatmospheric Temperatures," *Australian Journal of Physics* (1960), vol. 13, p. 299.

W. F. Hosford, *Mechanical Behavior of Materials*, 2nd ed., Cambridge University Press (2010).

W. F. Hosford and R. M. Caddell, *Metal Forming: Mechanics and Metallurgy*, 4th ed., Cambridge Univeristy Press (2011).

A. S. Keh and S. Weissmann, *Electron Microscopy and Strength of Crystals*, Interscience (1963).

W. C. Leslie, *The Physical Metallurgy of Steels*, Hemisphere (1981).

6

ANNEALING

GENERAL

Annealing is the heating of metal after it has been cold worked to soften it. Most of the energy expended in cold work is released as heat during the deformation. However, a small percent of the mechanical work is stored by dislocations and vacancies. The stored energy is the driving force for the changes during annealing. There are three stages of annealing. In order of increasing time and temperature, they are as follows:

1. *Recovery* – often a small drop in hardness and rearrangement of dislocations to form subgrains. Otherwise, overall grain shape and orientation remain unchanged. Residual stresses are relieved.
2. *Recrystallization* – replacement of cold-worked grains with new ones. There are new orientations, a new grain size, and a new grain shape, but not necessarily equiaxed. Recrystallization causes the major hardness decrease.
3. *Grain growth* – growth of recrystallized grains at the expense of other recrystallized grains.

RECOVERY

The energy release during recovery is largely due to annealing out of point defects and rearrangement of dislocations. Most of the increase

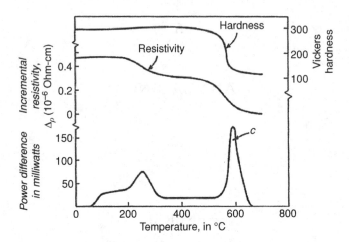

Figure 6.1. Property changes in tungsten during recovery. Note that the electrical con-ductivity improves before any major change in hardness. From A. Guy and J. Hren, *Elements of Physical Metallurgy*, 3rd ed., Addison-Wesley (1974).

of electrical resistivity during cold work is attributable to vacancies. These anneal out during recovery, so that the electrical resistivity drops (Figure 6.1) before any major hardness changes occur. During recovery, residual stresses are relieved, and this decreases the energy stored as elastic strains. The changes during recovery cause no changes in microstructure that would be observable under a light microscope. Figure 6.2 shows the energy release and the changes of resisitivity and hardness with increasing annealing temperatures.

Figure 6.2. Edge dislocations rearranging to form a low angle tilt boundary.

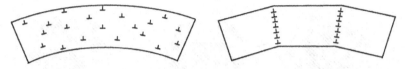

Figure 6.3. Alignment of low angle grain boundaries in a bent single crystal causes polygonization. From W. F. Hosford, *Physical Metallurgy*, 2nd ed., CRC Press (2010).

There is also a rearrangement of dislocations into lower energy configurations, such as low angle tilt boundaries (Figure 6.2). In single crystals, these boundaries lead to a condition called *polygonization*, in which a bent crystal becomes facetted (Figure 6.3). In polycrystals, these low angle boundaries form subgrains that differ in orientation by only a degree or two. Figure 6.4 shows the subgrains formed in iron.

There may be subgrain coalescence by rotation, as illustrated in Figure 6.5.

Figure 6.4. Subgrain formation in iron during recovery. From W. C. Leslie, J. T. Michalak, and F. W. Auk, Conference Series, *Iron and Its Dilute Solutions*, Interscience (1963).

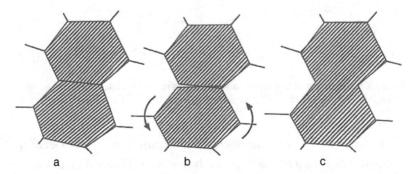

Figure 6.5. Subgrain coalescence by rotation. From W. F. Hosford, ibid.

In summary, during recovery, there is a large decrease in the number of point defects as they diffuse to edge dislocations and grain
boundaries. Dislocations rearrange themselves into lower energy configurations, leading to the formation of subgrains. Enough creep
occurs to reduce elastic strains. There are major changes in electrical properties, some softening, and a reduction of residual stresses,
but no major changes of microstructure.

RELIEF OF RESIDUAL STRESSES

To remove residual stresses, elastic strains must be converted to plastic strains. During a stress relief anneal, this occurs by creep. As the
creep occurs, the elastic strains decrease, lowering the driving force
for more creep and slowing the rate of stress relief. Figure 6.6 shows
how the rate of relief slows down. It also shows that stress relief is
faster at higher temperatures

RECRYSTALLIZATION

Recrystallization is a process in which old cold-worked grains are
replaced by new ones with many fewer dislocations as shown in
Figure 6.7. The new grains nucleate on the grain boundaries and junc-
tions of the old grains.

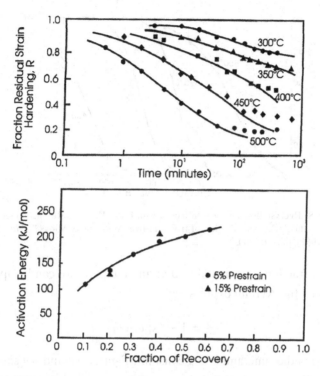

Figure 6.6. Effect of temperature and time on stress relief in iron strained 5%. From J. T. Mikalak and H. W. Paxton, *Trans. AIME*, v. 221 (1961).

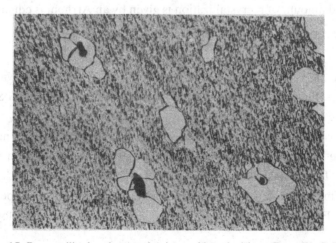

Figure 6.7. Recrystallized grains growing into cold-worked iron. From W. C. Leslie, J. T. Mikalak, and F. W. Aul, ibid.

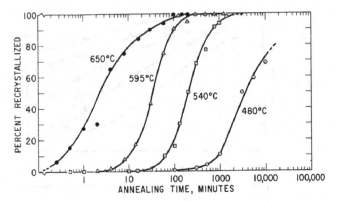

Figure 6.8. Recrystallization of iron 0.60% Mn. From W. C. Leslie, F. J. Plecity, and J. T. Mickalak, "Recrystallization of Iron and Iron-Manganese Alloys," *Trans. AIME*, v. 221 (1961), pp. 691–700.

The fraction, f, recrystallized at any temperature can be approximated by the Avrami expression:[*]

$$f = 1 - \exp(-bt), \tag{6.1}$$

where t is the time and b is a constant. Figures 6.8 and 6.9 show this progression.

Their growth depends on the temperature. The time, t, to achieve a fixed amount of recrystallization is given by an Arrhenius equation:

$$t = A\exp([Q/RT]), \tag{6.2}$$

where T is the absolute temperature and Q is the activation energy. Figure 6.10 shows that the recrystallization of Fe-3.5% Si in Figure 6.9 follows this equation.

It is common to define the recrystallization temperature of a metal as the temperature required to achieve 50% recrystallization in 1 hour. The recrystallization of pure metals is usually between 0.3 and 0.6 Tm, where Tm is the melting point on an absolute temperate scale. For iron this is between 520 and 959°C. Impurities and alloying

[*] M. Avrami, *Journal of Chemical Physics*, v. 7 (1939).

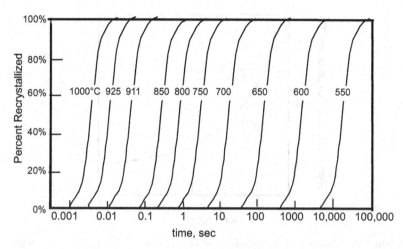

Figure 6.9. Recrystallization of Fe-3.5% Si after 60% cold work. Adapted from G. R. Speich and R. M. Fisher, *Recrystallization. Grain Growth and Textures*, ASM (1966).

elements increase the recrystallization temperature, and increased cold work decreases it, as shown in Figure 6.11.

Recovery before recrystallization has little effect on the recrystallization kinetics, as shown in Figure 6.12.

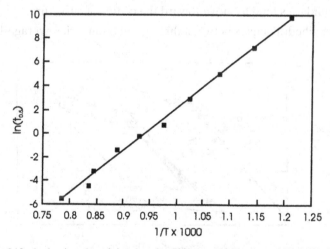

Figure 6.10. Arrhenius plot of the time for 50% recrystallization of Fe-3.5% Si using data in Figure 6.9.

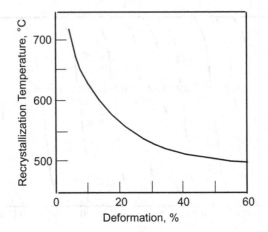

Figure 6.11. Lowering of recrystallization temperature of low-carbon steel with increased cold work. Data from H. F. Kaiser and H. F. Taylor, *Trans. ASM*, v. 27 (1939).

GRAIN GROWTH

In the grain-growth stage of annealing, recrystallized grains grow at the expense of other recrystallized grains. The driving force for grain growth is the reduction of energy associated with grain boundaries. Important principles include the following:

1. Boundaries tend to move toward the center of curvature.
2. The dihedral angles between three grain boundaries average 120°.

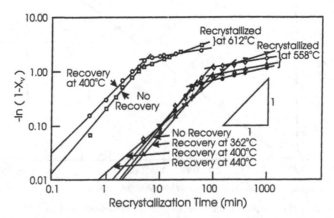

Figure 6.12. Recovery treatments prior to recrystallization has little effect. From A. Rosen, M. S. Burton, and G. V. Smith, *Trans. AIME*, v. 230 (1952).

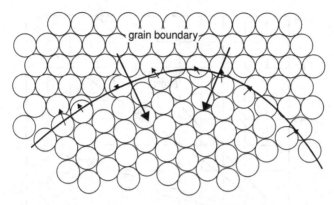

Figure 6.13. Movement of atoms to positions where they have a greater number of correct near neighbors. This is equivalent to the boundary moving toward its center of curvature. From W. F. Hosford, *Physical Metallurgy*, ibid.

Grain boundaries migrate toward their centers of curvature because by straightening, they reduce their surface area. This can also be understood from a microscopic viewpoint. Atoms tend move to positions with more correct near neighbors. This is equivalent to the boundary moving toward its center of curvature, as shown in Figure 6.13.

The consequence of these two principles can be understood best in terms of a two-dimensional model of grains, as illustrated schematically in Figure 6.14. If all of the grains were the same size, they would have six neighbors, and so the 120° requirement results in straight grain boundaries. A grain that is larger than its neighbors has

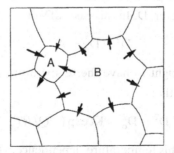

Figure 6.14. Large grains tend to be outwardly concave and therefore grow at the expense of their neighbors. Small grains tend to be inwardly concave and therefore shrink. From W. F. Hosford, ibid.

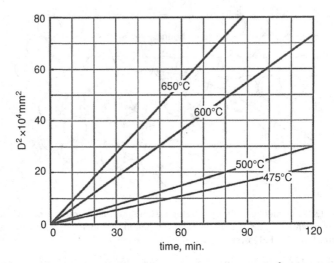

Figure 6.15. Grain growth of a 65% Cu–35% zinc brass. Note that D^2 is proportional to time as predicted by Equation 6.4. Data from P. Feltham and G. J. Copley, *Acta Met.*, v. 6 (1958). From W. F. Hosford ibid.

more than six neighbors, so its boundaries outwardly concave. Consequently, it will grow at the expense of its neighbors. A grain that is smaller than its neighbors has fewer than six neighbors, so its boundaries inwardly concave. It will shrink until it disappears. The result will be fewer grains, and so the average grain size will increase.

To the first approximation, one would expect that the rate of migration would be proportional to curvature $(1/r)$ and that the average curvature of the grain boundaries would be proportional to the average grain diameter, D. In this case, $dD/dt = a/D$, so

$$D^2 - D_0^2 = At. \tag{6.3}$$

The constant, A, ought to have the temperature dependence of the Arrhenius form, so that

$$D^2 - D_0^2 = Kt \exp\left(-Q/RT\right) \tag{6.4}$$

Figure 6.15 shows this temperature dependence. However, the kinetics is not often so well behaved.

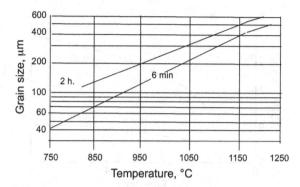

Figure 6.16. Increase of the austenite grain size of a 1060 steel with time and temperature. Data from O. O. Miller, *Trans. ASM*, v. 43 (1951).

Empirically the grain growth may be described by

$$D = kt^n, \qquad (6.5)$$

where n isn't necessarily 1/2. For iron, Hu[*] gives n = 0.4.

Figure 6.16 shows how the austenite grain size increases with time and temperature.

Alloying elements slow the rate of grain growth as illustrated in Figure 6.17.

Figure 6.18 shows that grain boundary migration rate depends on solute and misorientation. This orientation dependence is a major factor in the formation recrystallization textures.

Another reason is that second-phase particles tend to pin grain boundaries. Consider a grain boundary that is pinned by a spherical inclusion as shown in Figure 6.19. The contact length between the boundary and the inclusion is $2\pi r \cos \theta$, where r is the radius of the inclusion and θ is the contact angle. This creates a drag force normal to the inclusion of $2\pi r \gamma_{gb} \cos \theta$, where γ_{gb} is the grain boundary surface energy. The component normal to the grain boundry is $F = 2\pi r \gamma_{gb} \cos \theta \sin \theta$, which has a maximum value at $\theta = 45°$ of

$$F_{max} = \pi r \gamma_{gb}. \qquad (6.6)$$

[*] H. Hu, "Grain Growth in Zone-Refined Iron," *Canadian Metallurgy Quarterly*, v. 14 (1974).

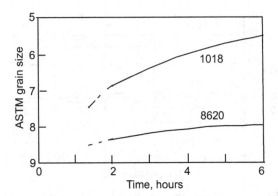

Figure 6.17. Effect of alloying on the rate of grain growth at 1010°C. The 8620 steel contains Cr, Mo, and Ni. Data from G. O. Ratliff and W. H. Samuelson, *Met. Prog.*, v. 108 (Sept. 1975).

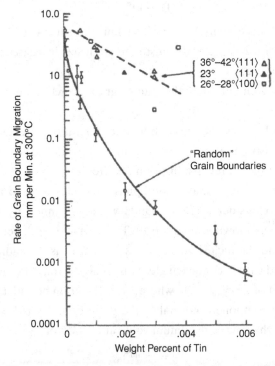

Figure 6.18. The rate of grain boundary migration in lead depends on the angle of misorientation as well as the amount of impurities. Special grain boundaries with certain angular relationships migrate much faster than normal grain boundaries. From J. W. Rutter and K. T. Aust. *Trans. AIME*, v. 23 (1919).

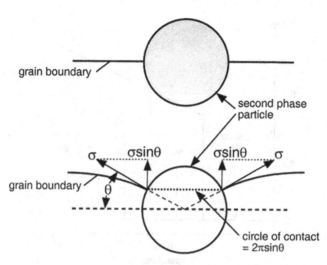

Figure 6.19. The drag of an inclusion on a grain boundary equals $\pi r \gamma$, where r is the particle radius and γ is the grain boundary surface energy. From W. F. Hosford, ibid.

The presence of inclusions can make a material resistant to grain growth. Such materials are called *inherently fine grained*. However, the resistance to grain growth may break down at sufficiently high temperatures, as shown in Figure 6.20. The result is *discontinuous grain growth or exaggerated grain growth* as illustrated in Figure 6.21. Dissolution of the second-phase particles or overcoming of their resistance allows a few grains to grow at the expense of others. This is sometimes called *secondary recrystallization* and can lead to the formation of still new crystallographic textures.

Grain growth is the growth of already-recrystallized grains at the expense of other recrystallized grains. The growth rate is retarded by impurities in solid solution, and those present as second-phase particles. There is a limiting grain size that decreases with more particles. If the particles dissolve or otherwise cease to prevent grain growth, there may be a sudden increase in grain size.

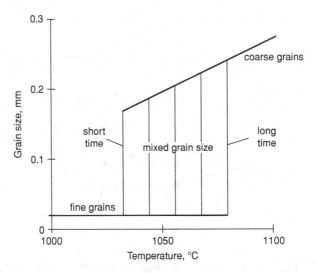

Figure 6.20. Inherently fine-grain steels (ones containing many fine particles) resist grain growth up to about 1050°C and then undergo exaggerated grain growth as the particle restrain is lost. Coarse grain steels (without many fine inclusions) undergo normal grain growth. From W. F. Hosford, ibid.

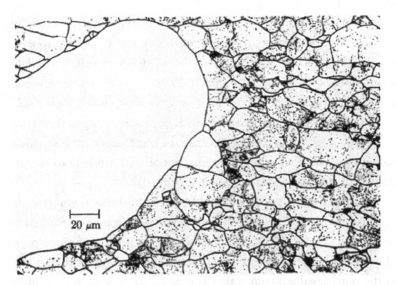

Figure 6.21. A large grain in a Fe-3% Si steel growing into finer-grained material. This is called *secondary recrystallization* or *exaggerated* or *discontinuous grain growth*. From A. G. Guy and J. J. Hren, *Elements of Physical Metallurgy*, 3rd ed., Addison-Wesley (1974).

REFERENCES

A. G. Guy and J. J. Hren, *Elements of Physical Metallurgy*, 3rd ed., Addison-Wesley (1974).

W. F. Hosford, *Physical Metallurgy*, 2nd ed., CRC Press (2010).

F. J. Humphreys and M. Hatherly, *Recrystallization and Related Annealing Phenomena*, Pergamon (1995).

G. Krauss, *Steels: Heat Treatment and Processing Principles*, ASM (1990).

W. C. Leslie, *The Physical Metallurgy of Steels*, Hemisphere (1981).

7

DEFORMATION MECHANISMS AND CRYSTALLOGRAPHIC TEXTURES

SLIP AND TWINNING SYSTEMS

The direction of slip in iron as in all body-centered cubic (bcc) metals is <111>. This is the direction of the shortest repeat distance in the bcc lattice. Slip can occur on any plane containing a <111> direction. The critical stress for slip is lowest for the {110} plane and highest for the {112}, the difference increasing as the temperature decreases. There is also an asymmetry to slip on {112}. The shear stress for slip is least when the direction is the same as for twinning and highest in the anti-twinning direction. See Figure 7.1.

For iron and all bcc metals, the twinning direction is <111>, and the twinning plane is {112}. The twinning shear strain is $1/\sqrt{2} = 0.707$. Twins, called *Neuman* bands, are very narrow and occur only during deformation at low temperatures or high strain rates. At 4.2 K, the critical shear stress for twinning is about 520 MPa. Twinning normally contributes little to the overall deformation.

WIRE TEXTURES IN BCC METALS

All bcc metals have similar deformation textures. The wire textures have <110> directions aligned with the wire axis. Peck and Thomas[*]

[*] J. F. Peck and D. A. Thomas, "A Study of Fibrous Tungsten and Iron," *Trans. TMS-AIME*, v. 221 (1961).

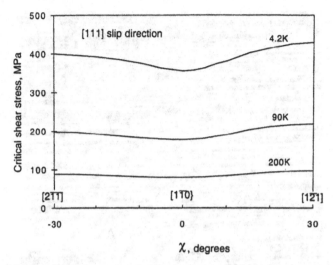

Figure 7.1. The critical shear stress for slip depends on the angle, χ, between <110> and the normal to the plane of maximum shear stress. $\chi = 0$ corresponds to {110}, and $\chi = 30°$ corresponds to {112}. From William F. Hosford in *Encyclopedia of Materials Science and Technology*, Elsevier Science (2001). Data from K. Kitajima, Y. Aono, and E. Kuramoto, *Scripta Metallurgica*, v. 13 (1979).

observed that the microstructures of drawn wires of tungsten metals contain grains that are elongated along the wire axis but seem to curl about one another, as shown in Figure 7.2. A similar microstructure is observed in draw iron wires (Figure 7.3).

These microstructures suggest that the grain in the drawn wires have the shape of long ribbons folded about the wire axis, as illustrated schematically in Figure 7.4.

The reason for this microstructure can be understood in terms of the <110> crystallographic texture in wires of bcc metals (Figure 7.5). With a <110> direction parallel to the wire axis, two of the four <111> slip directions allow thinning parallel to the <001> lateral direction but not to the <1$\bar{1}$0> lateral direction. The other two <111> slip directions are perpendicular to the wire axis and should not operate. The curly microstructure occurs because differently oriented grains contract in different directions and must curl about one

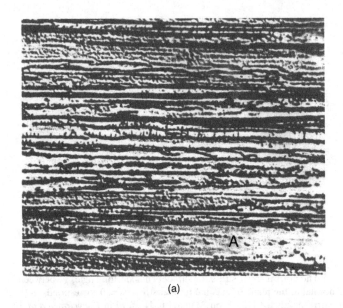

(a)

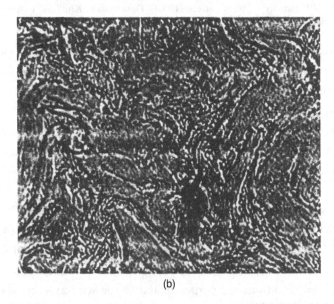

(b)

Figure 7.2. Microstructures of heavily drawn tungsten wires. (a) Longitudinal section with the wire axis horizontal. (b) Curly grains viewed parallel to the wire axis. From J. F. Peck and D. A. Thomas, "A Study of Fibrous Tungsten and Iron," *Trans. TMS-AIME*, v. 221 (1961).

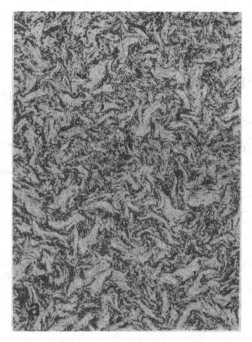

Figure 7.3. Curly grain structure of iron wire drawn to a strain of 2.7. From W. C. Leslie, *Physical Metallurgy of Steels*, Hemisphere (1981).

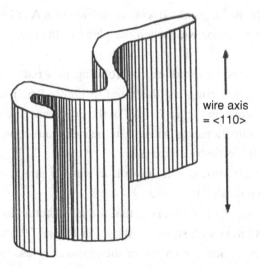

wire axis
= <110>

Figure 7.4. Schematic drawing of the shape of grains in drawn wires of bcc metals. From W. F. Hosford, *Mechanics of Crystals and Textured Polycrystals*, Oxford Science (1993).

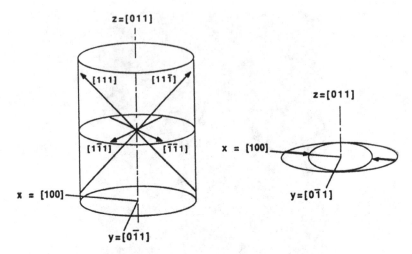

Figure 7.5. With a [011] direction parallel to the wire axis, the [111] and [11$\bar{1}$] slip directions cause thinning parallel to the lateral [100] direction without any thinning parallel to [0$\bar{1}$1]. The other two <111> slip directions are unstressed. From W. F. Hosford, ibid.

another to remain compatible. This explains the appearance of a few wide grains in the longitudinal section such as grain A in Figure 7.2(a). A grain appears to be wide when the plane of the section is parallel to [0$\bar{1}$1].

Swaged wires are produced by rotating dies reducing the wire diameter by hammering, as illustrated in Figure 7.6. This produces the microstructure in Figure 7.7. All of the grains are oriented so that their <100> directions are oriented in the radial direction, producing a <011><001> cylindrical texture.

The subgrain structure in drawn iron wires is similar to the grain structure, as indicated in Figure 7.8.

Langford and Cohen observed that the strain hardening caused by drawing was linear with strain, as shown in Figure 7.9. Probably this is because the critical dimension of the grains decrease linearly with elongation instead of with the square root of elongation.

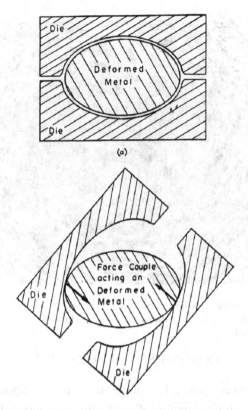

Figure 7.6. Swaging of a wire causes it to twist, producing the microstructure in Figure 7.7. From J. F. Peck and D. A. Thomas, ibid.

The microstructure of pearlite after wire drawing is similar to that of pure bcc metals (Figure 7.10). During wire drawing, the ferrite phase develops a <110> wire texture.

ROLLING TEXTURE

The rolling texture of iron (Figure 7.11) can be described as a <110> {11$\bar{2}$} texture, which means that a <110> direction aligned with a rolling direction and a {112} plane is the rolling plane normal. There is some rotation of this ideal orientation about the <110> rolling direction.

Figure 7.7. Cross-sectional microstrucrure of an iron wire swaged to an 87% reduction of area. Note that the thin [001] directions are aligned radially. From J. F. Peck and D. A. Thomas, ibid.

COMPRESSION TEXTURE

The compression texture is composed of two components. Some grains have a <100> direction aligned with the compression axis, and other grains have <111> so aligned.

The wire and compression textures can be understood roughly by the lattice rotations in single crystals undergoing <111> pencil glide. Figure 7.12 shows that two slip systems are active in different regions of the basic stereographic triangle.

In tension, the lattice rotates toward the slip direction, as shown in Figure 7.13. For both regions, the end texture is <110>. In compression, the lattice rotates toward the slip plane normal, which for pencil glide is equivalent to rotation away from the slip direction (Figure 7.14). Some of the grains end up with <100> parallel to the compression axis and others with <111> parallel to the compression axis.

Figure 7.8. Subgrains in iron wire drawn to a strain of 3.8. Longitudinal section top and cross-section bottom. From G. Lankford and M. Cohen, "Strain Hardening of Iron by Severe Plastic Deformation," *Trans. ASM*, v. 62 (1969).

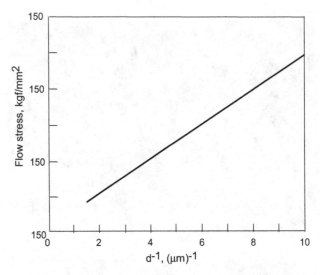

Figure 7.9. Increase of tensile strength of drawn wires with drawing strain for iron. Note the tensile strength increases linearly with $1/d$. From G. Lankford and M. Cohen, *Trans. ASM*, v. 62 (1969).

Figure 7.10. Scanning electron microscope microstructure of pearlite drawn to a strain of 3.2. From G Langford, *Met. Trans. A*, v. 8A (1977).

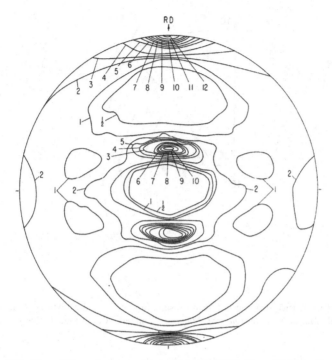

Figure 7.11. The <110> pole figure of iron cold rolled 90%. From W. C. Leslie, "Control of Annealing Texture by Precipitation in Cold-Rolled Iron," *Trans. AIME*, v. 221 (1961).

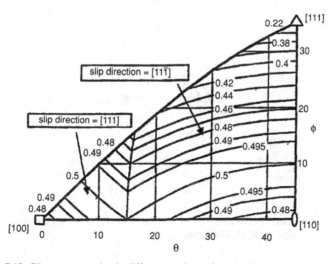

Figure 7.12. Slip systems active in different regions of the basic stereographic triangle. From W. F. Hosford, ibid.

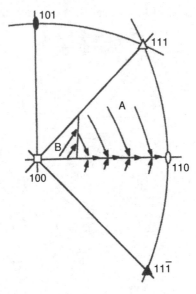

Figure 7.13. Lattice rotation in tension results in a <110> wire texture. From W. F. Hosford, *Mechanical Behavior of Materials*, 2nd ed., Cambridge University Press (2010).

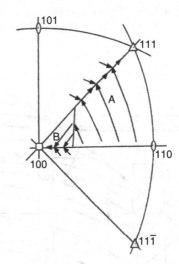

Figure 7.14. Lattice rotation causes either <100> or <111> to be aligned with the compression axis. From W. F. Hosford, ibid.

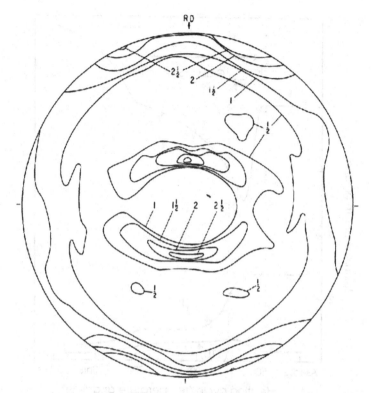

Figure 7.15. {110} pole figure of cold-rolled and recrystallized interstitial-free steel. From W. C. Leslie, ibid.

RECRYSTALLIZATION TEXTURES

The recrystallization texture of rolled low-carbon steel is character-ized by a strong {111} component in the plane of the rolled sheet. Figure 7.15 shows the {110} pole figure of interstitial-free steel recrys-tallized after an 80% cold-rolling reduction. The strong {111} compo-nent is indicated by the ring of high-intensity {110} poles about 35° from the center. The intensities at ± 30° to the rolling direction sug-gest that the texture can be described as <112>{11$\bar{1}$} with some rota-tion about the rolling direction. Figures 7.16 and 7.17 indicate that

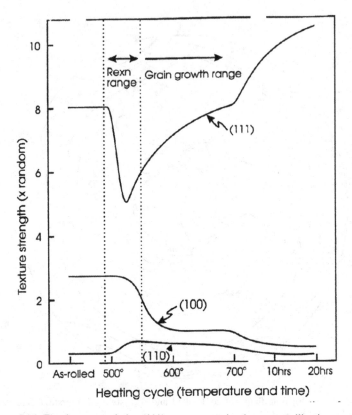

Figure 7.16. The increase of the {111} component in the recrystallization textures with increasing annealing temperatures. From W. B. Hutchinson, "Development and Control of Annealing Textures in Low Carbon Steel," *Int. Met. Rev.*, v. 29 (1984).

the strength of the {111} component increases with increasing rolling reductions and recrystallization temperatures.

Generally, a strong {111} texture is preferred for forming sheet steel because it increases the R-value and therefore sheet drawablity. It also and decreases the wrinkling tendency.

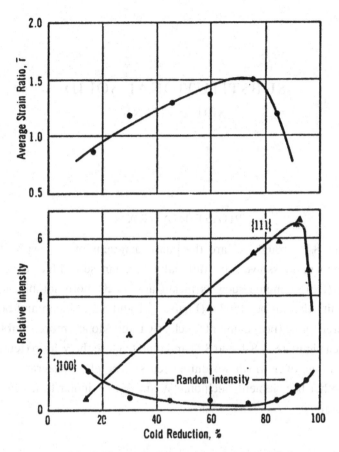

Figure 7.17. Dependence of recrystallization texture and plastic strain ratio, R, on the rolling reduction before recrystallization. From D. J. Blickwede, "Strain Hardening and Anisotropy," *Met. Prog.* (Jan. 1969).

REFERENCES

W. F. Hosford, *Mechanics of Crystals and Textured Polycrystals*, Oxford Science (1993).

W. F. Hosford, *Mechanical Behavior of Materials*, 2nd ed., Cambridge University Press (2010).

W. C. Leslie, *The Physical Metallurgy of Steels*, Hemisphere (1981).

G. Lankford in *Trans ASM,* v. 62 (1969).

8

SUBSTITUTIONAL SOLID
SOLUTIONS

PHASE DIAGRAMS

Figures 8.1 through 8.5 are the phase diagrams of iron with other elements that dissolve substitutionally. Manganese and face-centered cubic (fcc) elements such as nickel are much more soluble in fcc austenite than in bcc ferrite (Figures 8.1 and 8.2). Silicon and body-centered cubic (bcc) elements such as Cr and Mo are more soluble in ferrite (Figures 8.3, 8.4, and 8.5) and tend to close the γ-loop (temperature range over which austenite occurs). The phase diagrams of Fe with other bcc elements including W and V are similar to the Fe-Mo diagram.

TERNARY PHASE DIAGRAMS

Figures 8.6, 8.7, and 8.8 are sections of ternary iron-carbon diagrams with chromium, molybdenum, and tungsten.

EFFECTS OF SOLUTES ON THE EUTECTOID
TRANSFORMATION

Solute elements either raise or lower the Fe-C eutectoid temperature. Elements such as nickel and manganese that stabilize austenite lower the eutectoid temperature, whereas elements that stabilize

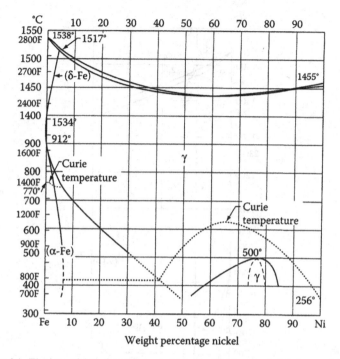

Figure 8.1. The iron-nickel phase diagram. Note that fcc iron and nickel are soluble in all proportions. From *Metals Handbook*, 8th ed., v. 8, ASM (1973).

ferrite raise it. All solutes lower the carbon content of the eutectoid composition (Figure 8.9). Note that carbon and nitrogen, which have greater solubilities in austenite than in ferrite, also lower the α–γ transition temperature.

EFFECT OF SOLUTES ON PHYSICAL PROPERTIES

Figure 8.10 shows the effects of solutes on the lattice parameter of ferrite. Figure 8.11 shows how solutes affect Young's modulus.

SOLID SOLUTION HARDENING

Figure 8.12 shows the solid solution hardening of ferrite by several alloying elements.

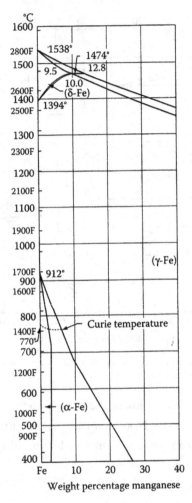

Figure 8.2. The iron-manganese phase diagram. From *Metals Handbook*, ibid.

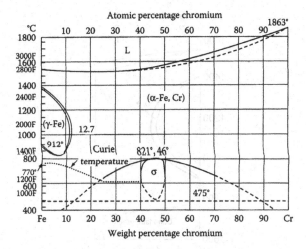

Figure 8.3. The iron-chromium phase diagram. Note that fcc iron and chromium are soluble in all proportions. From *Metals Handbook*, ibid.

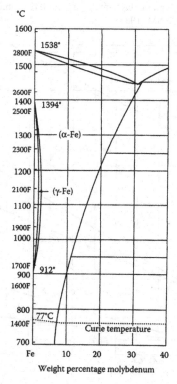

Figure 8.4. The iron-molybdenum phase diagram. From *Metals Handbook*, ibid.

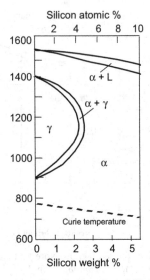

Figure 8.5. The iron-silicon phase diagram. Data from J. K. Stanley, *Electrical and Magnetic Properties of Metals*, ASM (1963).

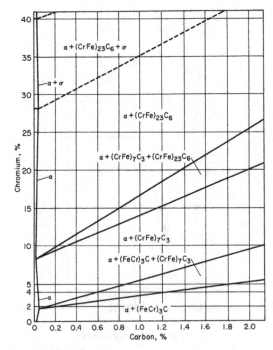

Figure 8.6. Fe-Cr-C ternary phase diagram. From E. G. Bain and H. W. Paxton, *Alloying Elements in Steel*, ASM (1939).

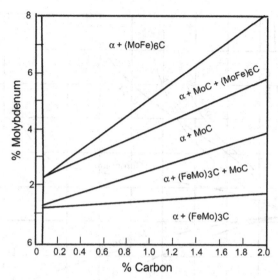

Figure 8.7. Fe-Mo-C ternary phase diagram. Data from E. G. Bain and H. W. Paxton, ibid.

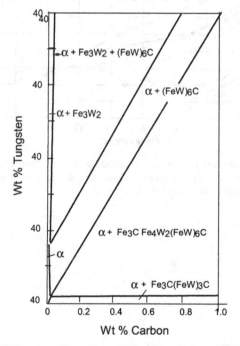

Figure 8.8. Fe-W-C ternary phase diagram. Data from Bain and H. W. Paxton, ibid.

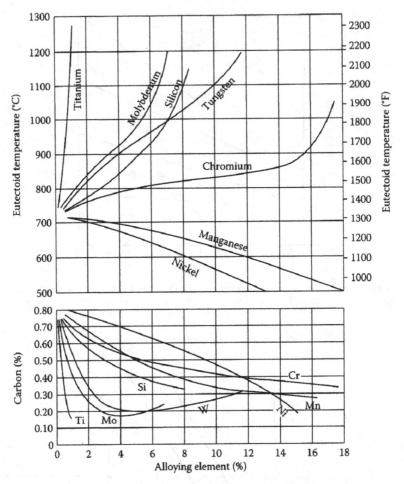

Figure 8.9. The eutectoid temperature is raised by ferrite stabilizers and lowered by austenite stabilizers. All alloying elements lower the carbon content of the eutectoid. From E. C. Bain and H. W. Paxton, ibid.

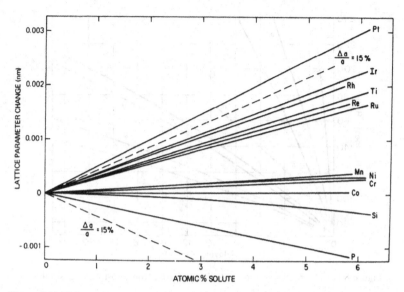

Figure 8.10. Effect of solutes on the lattice parameter of ferrite. From W. C. Leslie, *The Physical Metallurgy of Steels*, Hemisphere (1981).

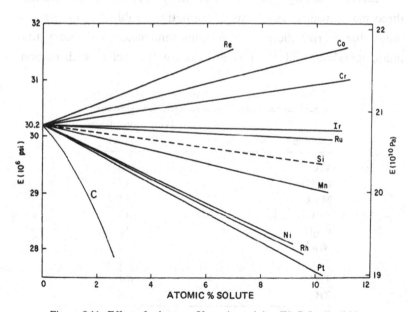

Figure 8.11. Effect of solutes on Young's modulus. W. C. Leslie, ibid.

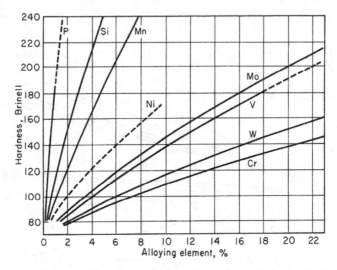

Figure 8.12. The solid solution hardening of ferrite from substitutional solutes.

CARBIDE-FORMING TENDENCIES

The carbide-forming tendencies of solutes in iron may be divided into three main groups. The heats of formation (Table 8.1) of the carbides characterize the carbide-forming tendencies. Vanadium, titanium, tantalum, and niobium are extremely reactive with carbon.

Table 8.1. *Heats of formation of metal carbides*

Carbide	Formation energy (kJ/mole)
V_4C	−49.3
Cr_7C3	−14.1
Mn_7C_6	−9.1
Fe_3C	−4.7
Co_2C	−2.8
Ni_3C	−1.2
TiC	−184
TaC	−162
ZrC	−208

These elements are used to remove dissolved carbon in stainless steel and interstitial-free steel. A second group, although not so strong, is stronger than iron, so during austenite transformation, carbon tends to segregate to the carbide phase rather than ferrite. In steels containing 10% Mn, the Mn/Fe ratio in the carbide is 1:4 instead of 1:10 in the steel as a whole. In a steel with 10% Cr and 1.0% C, the Cr/Fe ratio in the carbide is 4:6 instead of 1:10 in the steel as a whole. Other elements such as nickel, silicon, and copper do not form carbides in steel and segregate to the ferrite. Cobalt tends to distribute evenly between the carbide and ferrite and therefore has almost no effect on hardenability.

SOLUTE SEGREGATION TO GRAIN BOUNDARIES

All solutes tend to segregate at grain boundaries. The ratio of grain-boundary concentration to overall concentration increases with atomic solubilitiy.

REFERENCES

E. C. Bain and H. W. Paxton, *Alloying Elements in Steel*, ASM (1939).
W. C. Leslie, *The Physical Metallurgy of Steels*, Hemisphere (1981).
The Making, Shaping and Treating of Steel, U.S. Steel Corp. (1971).
Metals Handbook, 8th ed., v. 8, ASM (1973).

9

INTERSTITIAL SOLID SOLUTIONS

ATOMIC DIAMETERS

The only atoms small enough to dissolve interstitially in iron are hydrogen, boron, nitrogen, and carbon. The atomic diameters of these elements and iron are given in Table 9.1.

LATTICE SITES FOR INTERSTITIALS

Figure 9.1 shows the largest holes for interstitial site in bcc-ferrite lattice. It is often claimed that the interstitial atoms occupy sites in the centers of the faces and of the edges. In this case, the ratio of the size of the hole to that size of the lattice atoms would be

$$d/D = 2\sqrt{3}/3 - 1 = 0.154. \tag{9.1}$$

However, there are larger sites coordinated between four lattice atoms, as illustrated in Figure 9.1. The ratio of the size of these holes to that size of the lattice atoms is

$$d/D = \sqrt{(5/3)} - 1 = 0.291. \tag{9.2}$$

This site is about 90% larger. There is evidence that hydrogen occupies the larger site, whereas calculations suggest that carbon

Table 9.1. *Atomic diameters*

Element	Diameter (nm)
H	0.092
B	0.092
N	0.142
C	0.154
α-Fe	0.2482
γ-Fe	0.2538

occupies the smaller site. In either case, interstitial carbon and nitrogen atoms are much too large to fit in these holes, so solubility is limited.

The largest holes for interstitial sites in face-centered cubic (fcc) lattice are at the center of the unit cell and in the centers of the edges, as shown in Figure 9.2. The ratio of the size of the hole to that size of the lattice atoms is

$$d/D = \sqrt{2} - 1 = 0.414. \qquad (9.3)$$

This is much larger than the hole in the body-centered cubic (bcc) lattice, so the solubilities of carbon and nitrogen are much larger in austenite than in ferrite. However, it is still smaller that the ratio of

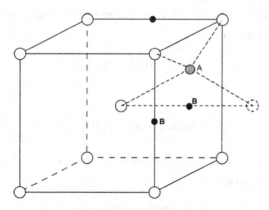

Figure 9.1. Interstitial site in the bcc lattice.

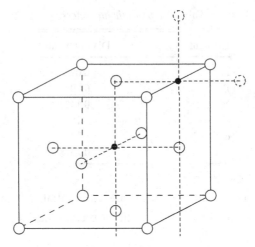

Figure 9.2. Interstitial sites in the fcc lattice.

the diameters of interstitial atoms to iron atoms, so the solubility is limited.

Except hydrogen, all interstitially dissolved atoms are larger than the holes between iron atoms in the lattice, so they cause volume expansion.

LATTICE EXPANSION WITH C, N

At room temperature, the effects of carbon and nitrogen on the lattice parameters in nm are

$$a = 0.2866 + 0.84 \times 10^{-3} \ (\%C) \tag{9.4}$$

and

$$a = 0.2866 + 0.79 \times 10^{-3} \ (\%N). \tag{9.5}$$

Carbon dissolved in ferrite makes the bcc structure slightly tetragonal. The c/a ratio is given by

$$c/a = 1.000 + 0.045 \ (\%C) \tag{9.6}$$

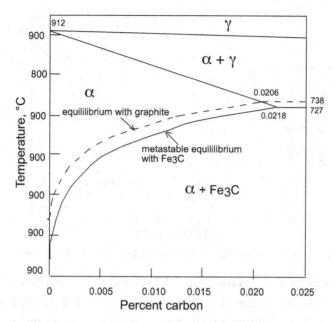

Figure 9.3. Solubility of carbon in ferrite. Data from J. Chipman, *Metals Handbook*, 8th ed., v. 8, ASM (1973).

SOLUBILITY OF CARBON AND NITROGEN

The solubility of carbon in α-iron in equilibrium with graphite is given by

$$\log_{10}[C]_{ppm} = 7.81 - 5550/T. \tag{9.7}$$

For metastable equilibrium with Fe_3C, the solubility is

$$\log_{10}[C]_{ppm} = 6.38 - 4040/T \tag{9.8}$$

and for metastable equilibrium with ε-carbide the solubility is

$$\log_{10}[C]_{ppm} = 4.06 - 1335/T. \tag{9.9}$$

These are plotted in Figure 9.3. The solubility of carbon in γ-iron and δ-iron are shown in Figures 9.4 and 9.5.

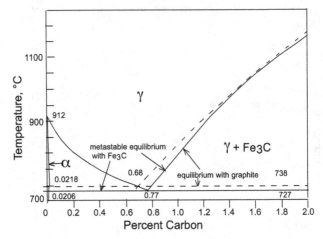

Figure 9.4. Solubility of carbon in austenite. Data from J. Chipman, ibid.

Comparison of Figures 9.4 and 9.5 show that the solubility of carbon in γ-iron is much greater than in ferrite. Again, the solubility with respect to the equilibrium with graphite is lower than for metastable equilibrium with Fe₃C.

The solubility of nitrogen in α-ferrite is plotted in Figure 9.6.

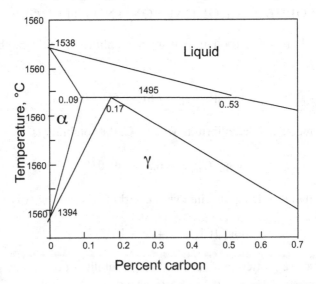

Figure 9.5. Solubility of carbon in delta ferrite. From J. Chipman, ibid.

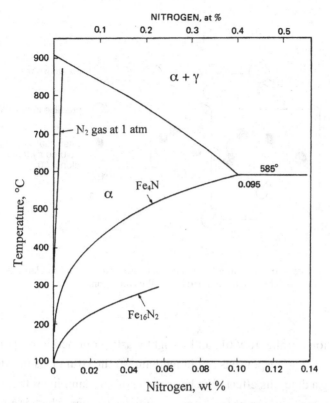

Figure 9.6. Solubility of nitrogen in ferrite. From J. D. Fast and M. B. Verrijp, "Solubility of Nitrogen in Iron," *J. Iron Steel Inst.*, v. 180 (1955).

SNOEK EFFECT IN BCC METALS

The positions of the interstitial atoms in bcc metals occupy positions that are illustrated in Figure 9.7. When the lattice is under no external force, all of these sites are equivalent, each interstitial atom being midway between two iron atoms. However, with elastic extension along [001], one-third of the sites become more favorable. The Poisson contraction along [100] and [010] makes the other two-thirds of the sites less favorable. Given sufficient thermal energy and time, the interstitial atoms will jump to the favorable sites, causing a slight additional

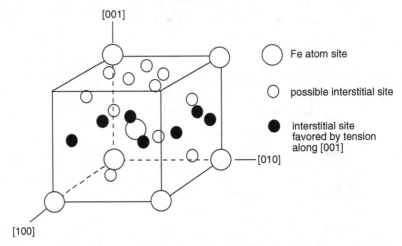

Figure 9.7. Interstitial sites in a bcc lattice. The sites indicated by the large dots are more favored when the cell is extended in the [001] direction.

extension parallel to [001] and a slight elastic contraction perpendicular to it. The response is not immediate, so the strain lags the stress, causing a damping effect. The basic cause of this damping is the jumping of interstitial atoms from one position to another, which is equivalent to diffusion. The damping is therefore frequency and temperature dependent with the frequency f^* at maximum damping being given by an Arrhenius relation,

$$f^* = f_o \exp[(-Q/R)/T], \tag{9.10}$$

where the activation energy, Q, is the same as that for diffusion of the interstitial atom. Figure 9.8 shows that as the frequency is increased, the damping peak occurs at higher temperatures.

Boron is of little concern except in hardenable steels (see Chapter 12).

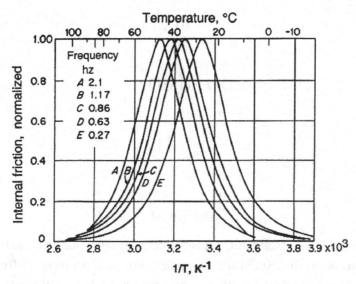

Figure 9.8. Damping of iron as a function of frequency and temperature. From C. Wert and C. Zener, "Interstitial Atomic Diffusion Coefficients," *Phys. Rev.*, v. 76 (1949). Copyright 1949 by American Physical Society.

REFERENCES

W. F. Hosford, *Physical Metallurgy*, 2nd ed., CRC Press (2010).
W. C. Leslie, *The Physical Metallurgy of Steels*, Hemisphere (1981).
Metals Handbook, 8th ed., v. 8, ASM (1973).

10

DIFFUSION

GENERAL

In interstial solid solutions, diffusion occurs by interstital atoms jumping from one interstitial site to another. For an atom to move from one interstitial site to another, it must pass through a position where its potential energy is a maximum. The difference between the potential energy in this position and that in the normal interstitial site is the activation energy for diffusion and must be provided by thermal fluctuations. The overall diffusion rate is governed by an Arrhenius-type rate equation,

$$D = D_o \exp(-E/kT), \tag{10.1}$$

where D_o is a constant for the diffusing system, k is Boltzmann's constant, T is the absolute temperature, and E is the activation energy (the energy for a single jump). Often this equation is written as

$$D = D_o \exp(-Q/RT), \tag{10.2}$$

where the activation energy, $Q = N_o E$, is for a mole of jumps. (N_o is Avogadro's number $= 6.02 \times 10^{23}$ jumps/mole). Correspondingly, R is the gas constant ($N_o k = 4.18$ J/mole).

The rates of interstitial diffusion in face-centered cubic (fcc) metal are much higher than in body-centered cubic metals. This is because the passageways between equivalent interstitial sites is larger in the fcc

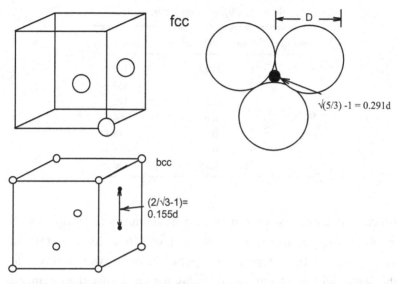

Figure 10.1. Diffusion paths for interstitials in fcc and bcc lattices. Note that the path in the bcc lattice is much more restricted than in the fcc lattice.

structure (0.291D for bcc vs. 0.155d for bcc, where d is the diameter of the solvent metal; see Figure 10.1).

Table 10.1 shows the values of Q and d_o for diffusion in α- and γ-iron. Experimental data for diffusion of interstitials in several metals are given in Table 10.2.

MECHANISMS OF DIFFUSION

Interstitial solutes diffuse much faster than substitutional solutes. Table 10.2 shows that carbon, which dissolves interstitially, diffuses much faster than do nickel and manganese, which dissolve substitutionally. Note also that diffusion in bcc iron is much faster than in fcc iron.

For diffusion in substitutional solid solutions and self-diffusion, the diffusion mechanism is not so obvious. For years, many experts considered that diffusion occurs by an interchange mechanisms (in

Table 10.1. *Diffusivities for interstitials*

Solvent	Solute	D_o(m^2/s)	Q(kJ/mole)
Ta	O	0.44×10^{-6}	107
	N	0.56	159
	C	0.61	162
Fe	C	2.0	84.4
	N	0.3	76.4
	H	0.1	13.4
Ni	H	0.45	36.1

which two atoms exchanged places) and ring mechanisms (which involves a cooperative rotation of a ring of 4, 6, or more atoms). In mechanisms of these types, both species of atoms would diffuse at the same rate. See Figure 10.1. Today we know that the dominant mechanism of diffusion is by movement of vacancies (vacant lattice

Table 10.2. *Diffusion Constants for Several Systems*[*]

Solute	Solvent	Solvent structure	D_o(m^2/s)	Q(kJ/mole)
Carbon	α-iron	bcc	2.2×10^{-4}	122
Carbon	γ-iron	fcc	0.15×10^{-4}	142
Nitrogen	α-iron	bcc	8×10^{-7}	76
Hydrogen	α-iron	bcc	2.3×10^{-9}	6.6
Nickel	α-iron	bcc	9.9×10^{-4}	259
Manganese	α-iron	bcc	4.9×10^{-5}	276
Molybdenum	α-iron	bcc	1.5×10^{-2}	283
Tungsten	α-iron	bcc	6.9×10^{-3}	265
Chromium	α-iron	bcc	2.5×10^{-4}	240
Chromium	γ-iron	fcc	2.2×10^{-4}	268
Copper	γ-iron	fcc	3.0×10^{-4}	255
Manganese	γ-iron	fcc	0.35×10^{-4}	282

[*] Most of the data are from a compilation in D. S. Wilkinson, *Mass Transport in Solids and Fluids*, Cambridge Solid State Series (2000).

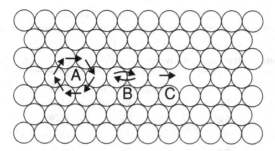

Figure 10.2. Schematic illustration of several mechanisms proposed for substitutional diffusion. (A) Ring interchange. (B) Simple interchange. (C) Vacancy migration. From W. F. Hosford, *Physical Metallurgy*, 2nd ed., CRC Press (2010).

sites). The equilibrium number of vacancies depends exponentially on temperature:

$$n_v = n_o \exp(-E_f/kT), \tag{10.3}$$

where n_v/n_o is the fraction of the lattice sites that are vacant and E_f is the energy to form a vacancy. The rate a given vacant site will be filled by a substitutional atom moving into it is also dependent on thermal activation,

$$\text{rate} = \exp(-E_m/kT), \tag{10.4}$$

where E_m is the energy barrier to the movement of a vacancy by its being filled by an adjacent substitutional atom. The net rate of diffusion is proportional to the product of the number of vacancies and the rate at which they contribute to diffusion. Therefore, $D = D_o \exp(-E_f/kT) \cdot \exp(-E_m/kT)$, which simplifies to

$$D = D_o \exp(-E/kT), \tag{10.5}$$

where

$$E = E_f + E_m. \tag{10.6}$$

Table 10.3. *Self-diffusion data*

Metal	Crystal structure	Q kJ/mole	D_o (m^2/s)	T_m (K)	Q/RT_m
Cu	fcc	198	20×10^{-6}	1356	17.5
Ag	fcc	185	40	1234	18.0
Ni	fcc	281	130	1726	19.6
Au	fcc	198	9.1	1336	17.8
Pb	fcc	102	28	600	20.4
α-Fe	bcc	240	190	1809	15.9
Nb	bcc	441	1200	2741	19.4
Mo	bcc	461	180	2883	19.2
Mg	hcp	136	125	923	17.8

hcp = hexagonal close-packed

Of course, this equation can also be expressed in an equivalent form in terms of Q, the activation energy per mole of diffusion jumps,

$$D = D_o \exp(-Q/RT). \qquad (10.7)$$

In general, the activation energies for self-diffusion and diffusion of substitutional solutes are considerably higher than those for interstitial diffusion, and therefore the diffusion rates are much lower. Data for self-diffusion in several metals are given Table 10.3. In comparing these data, several other trends are apparent. One is that the activation energies increase with melting point. In fact, for most relatively close-packed metals (fcc, bcc, hexagonal close packed), Q/T_m is nearly the same.

DIFFUSION OF INTERSTITIALS

Figure 10.3 shows that the diffusivities of interstitially dissolved H, N, and C are much greater than elements that are substitutionally dissolved elements.

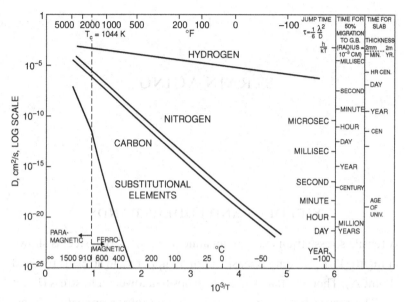

Figure 10.3. Diffusivity of interstitials. From W. C. Leslie, *The Physical Metallurgy of Steels*, Hemisphere (1981).

REFERENCES

W. F. Hosford, *Physical Metallurgy*, 2nd ed., CRC Press (2010).

W. C. Leslie, *The Physical Metallurgy of Steels*, Hemisphere (1981).

J. D. Verhoeven, *Steel Metallurgy for the Nonmetallurgist*, ASM (2007).

D. S. Wilkinson, *Mass Transport in Solids and Fluids*, Cambridge Solid State Series (2000).

11

STRAIN AGING

YIELDING AND LÜDERS BANDS

A tensile stress-strain curve of an annealed low-carbon steel is shown in Figure 11.1. Loading is elastic until an upper yield stress is reached (Point A). Then the load suddenly drops to a lower yield stress (Point B). The reason for this phenomenon is that during annealing, interstitially dissolved carbon and nitrogen atoms tend to diffuse to edge dislocations. Because they partially relieve the stress field around the dislocations (Figure 11.2), they lower the energy. A higher stress is required to break the dislocations free from the interstitial atoms than to move them once they have broken free. Continued elongation after initial yielding occurs by the propagation of the yielded region at this lower yield stress until the entire gauge section has yielded (point C). During this period of yielding, there is a sharp boundary or *Lüders band* between the region that has yielded and the region that has not. Behind this boundary, all of the material has suffered the same strain. The Lüders strain or yield point elongation in low-carbon steels is typically 1 to 5%. Strain hardening starts only after the Lüders band has traversed the entire gauge section. Finally, at some point, D, the specimen undergoes necking. Figure 11.3 shows Lüders bands that have partially traversed the gauge section.

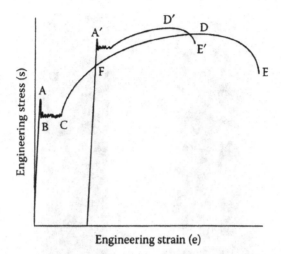

Figure 11.1. Tensile stress-strain curve of a low-carbon steel. From W. F. Hosford, *Physical Metallurgy*, 2nd ed., CRC Press (2010).

STRAIN AGING

If the specimen were unloaded at some point, F, and immediately reloaded, the stress-strain curve would follow the original curve. However, if the unloaded specimen were allowed to *strain age*, a new yield point would develop (A′). Strain aging would occur if the time was long enough and the temperature were high enough to allow

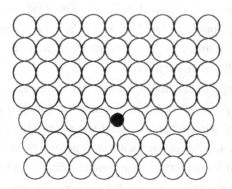

Figure 11.2. A small interstitial atom (black dot) helps relieve the hydrostatic tension at an edge dislocation.

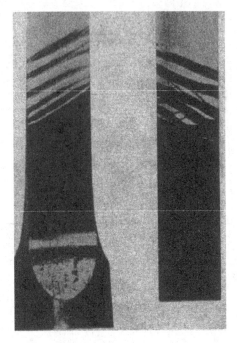

Figure 11.3. Lüders bands. Where they stop and start again leaves surface discontinuities. From F. Körber, *J. Inst. Metals*, v. 48 (1932).

interstitial carbon and nitrogen atoms to diffuse to the dislocations. Figure 11.4 shows that the amount of strain aging increases with the time at room temperature between unloading and reloading.

Strain aging is also time dependent. *The Making, Shaping and Treating of Steel,* published by U.S. Steel Corporation, gives the combinations of aging times and temperatures that cause the same amount of strain aging listed in Table 11.1.

A simple experiment that demonstrates the yield point effect can be made with piece of annealed florist wire, which is a low-carbon steel. When the wire is bent, it will form sharp kinks because once yielding occurs at a particular location, it takes less force to continue the bend at that location than to initiate bending somewhere else. On the other hand, copper wire, which has no yield point, will bend in a continuous arc (see Figure 11.5).

Table 11.1. *Combinations of aging times and temperatures*
that result in the same amount of strain aging

Temperatures (°C)	0	21	100	120	150
	1 yr	6 mo	4 hr	1 hr	10 min
	6 mo	3 mo	2 hr	30 min	5 min
	3 mo	6 wk	1 hr	15 min	2.5 min

The extent of the Lüders strain decreases with larger grain sizes
(see Figure 11.6).

Strain aging is generally considered undesirable in later forming
of sheets because it can lead to the formation of *stretcher strains*,
which are the result of nonuniform running of Lüders bands. Stretcher
strains (Figure 11.7) mar the surface appearance of finished parts.

Yield points and the occurrence of stretcher strains can be eliminated by either *roller leveling* or by *temper rolling*. In roller leveling,

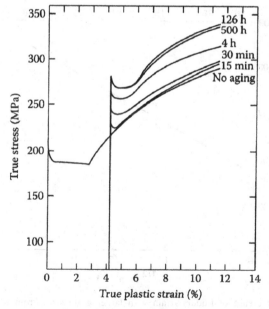

Figure 11.4. Effect of aging time and temperature in strain aging. From *Making, Shaping and Treating of Steel*, 9th ed., U. S. Steel Corp. (1970).

Figure 11.5. Bending of a low-carbon steel wire will result in kinks rather than a continuous curve as with cooper wire because of the tendency to localize deformation.

a sheet is passed between a set of rolls that flatten the sheet by bending and unbending (Figure 11.8). During temper rolling, also called *pinch pass*, the reduction is small. The strains in both processes are of the order of 0.5%, so the entire cross-section does not undergo plastic deformation (Figure 11.9).

Aluminum killing of steels deceases the susceptibility to strain aging. It should be realized that strain aging after forming is desirable, because it increases the strength. *Bake-hardenable* steels are formulated to strain age during paint baking.

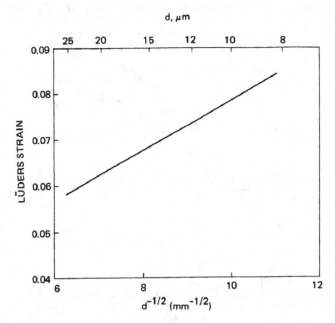

Figure 11.6. Decrease of Lüders strain with larger grain size. From R. W. Evans, "Effect of Prior Cold Reduction on the Ductility of Annealed Rimmed Steel," *J. Iron Steel Inst.*, v. 205 (1967).

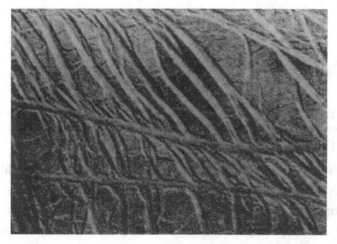

Figure 11.7. Stretcher strains resulting from incomplete running of Lüders bands. From *Metals Handbook*, 8th ed., v. 7, American Society for Metals (1962).

DYNAMIC STRAIN AGING

At somewhat elevated temperatures, strain aging can occur during deformation. In this case, it is called *dynamic strain aging*. Dynamic strain aging is manifested by serrated stress-strain curves and stretcher strains. The stress-strain curves are characterized by sudden load drops that correspond to bursts of plastic deformation followed by load drops and elastic reloading until another burst of deformation occurs.

Dynamic strain aging is the result of the attraction of solute atoms to dislocations. It is like static strain aging in that it takes a greater force to initiate plastic deformation by dislocations breaking free of

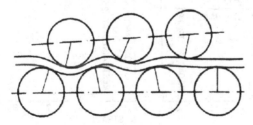

Figure 11.8. Roller leveling.

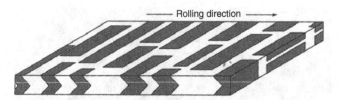

Figure 11.9. The very light reductions during temper rolling cause only small regions to deform plastically. From R. D. Butler and D. V. Wilson, "Mechanical Behaviour of Temper Rolled Steel," *J. Iron Steel Inst.*, v. 201 (1963).

solutes than to cause them to move once they are free. The difference is that dynamic strain aging occurs at a temperature and a strain rate that allows the solute atoms to pin the dislocations again while the material is deforming. The result is that the stress-strain curve consists of a series of load drops occurring during the deformation followed by elastic reloading until the stress is high enough to reinitiate deformation. Figure 11.10 shows the difference between static and dynamic strain aging in a low-carbon steel.

For low-carbon steels, static strain from interstitial solutes (C and N) aging occurs in the temperature range of 100 to 300°C, as shown in Figure 11.11. It may occur at lower temperatures if the amount of nitrogen in solution is high.

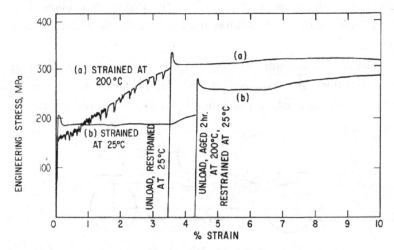

Figure 11.10. Dynamic and static stain aging of a 0.03% C rimmed steel. From A. S. Keh, in *Materials Sci. Res.*, v. 1 (1963).

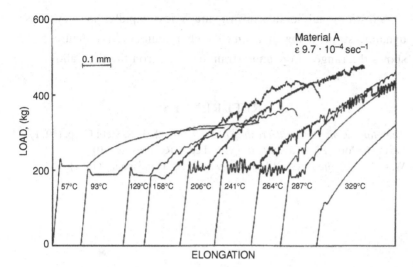

Figure 11.11. The temperature and strain-rate dependence for the onset of dynamic strain aging in steels containing three levels of nitrogen. From A. S. Keh, Y. Nakada, and W. C. Leslie, in *Dislocation Dynamics*, McGraw-Hill (1968).

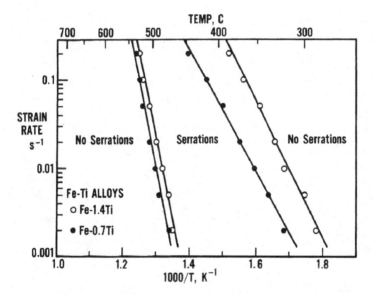

Figure 11.12. The temperature and strain-rate dependence of dynamic strain aging in iron-titanium alloys. From W. C. Leslie, L. J. Cuddy, and R. J. Sober, *Proc. 3rd Intern. Conf. Strength of Metals and Alloys*, v. 1, Institute of Metals, London (1973).

Substitutional solutes diffuse much less rapidly, so they cause dynamic strain aging at a much higher temperature. Figure 11.12 shows the range of dynamic strain aging in iron-titanium alloys.

REFERENCES

The Making, Shaping and Treating of Steel, 9th ed., U.S. Steel Corp. (1971).
W. F. Hosford, *Physical Metallurgy*, 2nd ed. CRC Press (2010).
W. C. Leslie, *The Physical Metallurgy of Steels*, Hemisphere (1981).

12

AUSTENITE TRANSFORMATION

KINETICS OF AUSTENIZATION

A plain-carbon steel contains ferrite and pearlite. To form a homogeneous solution of carbon in austenite requires heating for a long enough period of time for the carbon atoms to diffuse from the pearlite to the centers of the ferrite regions. The time required depends on the austenitizing temperature and the diffusion distance, as shown in Figure 12.1. These times are so short that for most purposes one can assume that autenitization is instantaneous.

Substitutionally dissolved elements diffuse much slower than carbon, so it often takes a long time to form a homogeneous solution. The time required for chromium to diffuse several distances is shown in Figure 12.2.

Figure 12.3 shows the effect of the austenitizing temperature on the rate of austenite formation.

PEARLITE FORMATION

The transformation of austenite to pearlite involves both nucleation of pearlite colonies and their growth. There is an activation barrier to nucleation of pearlite in austenite. The volume misfit strain, the new ferrite-cementite boundaries, and the new boundaries between austenite and the new phases increase the energy. Driving the reaction

113

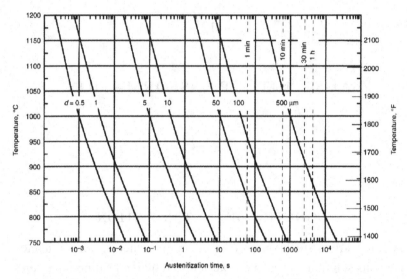

Figure 12.1. Temperature and times for a carbon atom to diffuse several distances. From J. D. Verhoeven, *Steel Metallurgy for the Nonmetallurgist*, ASM (2007).

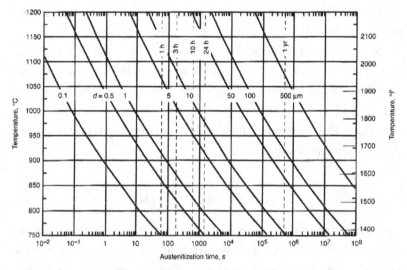

Figure 12.2. Temperature and times for chromium to diffuse several distances. From J. D. Verhoeven, ibid.

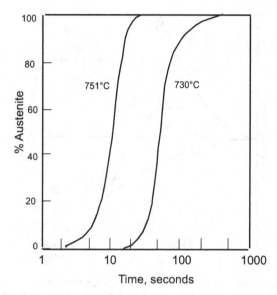

Figure 12.3. The effect of the austenitizing temperature on the rate of austenitization. Data from M. A. Grossman and E. C. Bain, *Principles of Heat Treatment*, ASM (1964).

is the change in volume free energy, ΔG_v, which is proportional to the undercooling. The loss of austenite grain boundaries lowers the activation barrier. Figure 12.4 shows these energy terms.

The growth of a pearlite colony into austenite is shown in Figure 12.5. As shown schematically in Figure 12.6, for a plain-carbon steel, growth depends on the diffusion of carbon from ahead of the growing ferrite to ahead of the growing cementite.

The transformation to pearlite involves both nucleation and growth. Both the nucleation rate and the growth rate increase with lower temperature, but the nucleation rate increases more rapidly (Figure 12.7). Therefore, the pearlite spacing becomes finer as the transformation temperature decreases.

As the pearlite forms, alloying elements must partition to either the ferrite or the cementite. Nickel and silicon partition to the ferrite, whereas carbide forming elements such as Cr, W, Mn, and Mo partition to the cementite. Figure 12.8 shows this partitioning.

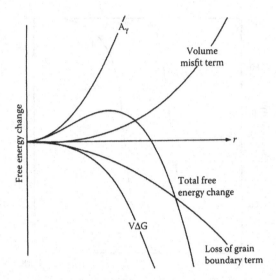

Figure 12.4. The volume misfit and new surface energy terms increase the activation energy for pearlite formation, whereas the loss of austenite grain boundary energy and the decrease of volume free energy tend to lower it.

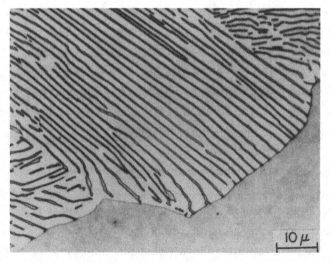

Figure 12.5. A pearlite colony growing into austenite. From J. R Vilella, in *Decomposition of Austenite by Diffusional Processes*, V. F. Zackay and I. Aaronson, eds, Interscience Publishers (1962).

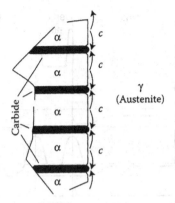

Figure 12.6. The growth of pearlite requires diffusion of carbon atoms in front of the growing ferrite to the growing cementite.

Because alloying elements diffuse more slowly than carbon, they slow the growth rate of pearlite (Figures 12.9 and 12.10).

As a steel is cooled below a critical temperature, M_s, it transforms rapidly by shear to a new crystal structure, *martensite*. Diffusion of carbon is not involved, so the amount of this transformation does

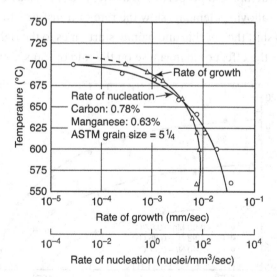

Figure 12.7. Increase of nucleation and growth rates with falling temperatures in a eutectoid steel. From R. F. Mehl and W. Hagel, originally in *Progress in Metal Physics*, v. 6 (1963).

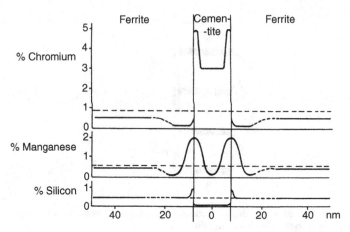

Figure 12.8. Partitioning of Cr, Mn, and Si between the ferrite and cementite phases. Note the segregation to the boundaries. From P. R. Williams, M. K. Miller, P. A. Beaven, and G. D. W. Smith, "Fine Scale Partitioning in Pearlite Slab. Transmission Using Electron Microscopy," *Phase Transformations*, v. 2, Institute of Metallurgists (1979), pp. II-93 to II-103.

not increase appreciably with time but depends only on the temperature. The structure and properties of martensite are discussed later.

Because alloying elements slow the rate of transformation of austenite, they shift the pearlite and bainite start times to the right. Figure 12.11 shows the effect of manganese on the rate of transformation.

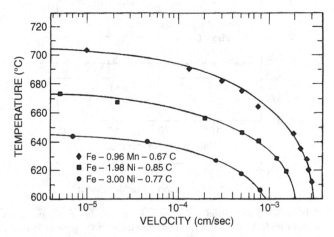

Figure 12.9. Effect of nickel and manganese on the growth rate of pearlite. From M. P. Puls and J. S. Kirkaldy, "Pearlite Reaction," *Met. Trans.*, v. 3 (1972).

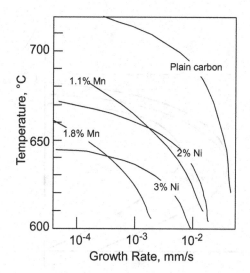

Figure 12.10. Alloying elements slow the growth rate of pearlite. Data from N. A. Razik, G. W. Lorimer, and N. Ridley, "An Investigation of Manganese Partitioning During the Austenite–Pearlite Transformation Using Analytical Electron Microscopy," *Acta Met.*, v. 22 (1974).

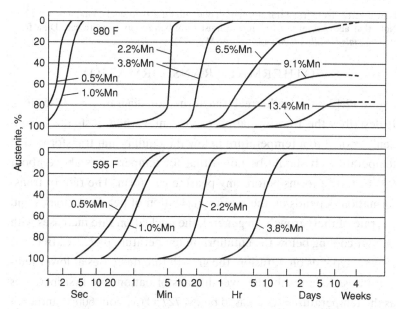

Figure 12.11. Effect of manganese on slowing transformation to ferrite and pearlite. From E. G. Bain and H. W. Paxton, *Alloying Elements in Steel*, ASM (1939).

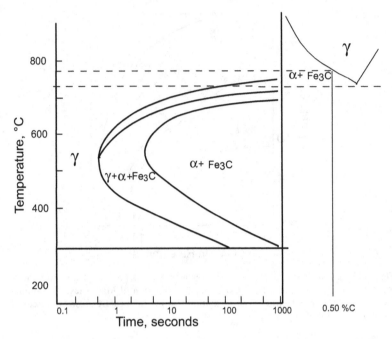

Figure 12.12. Isothermal transformation diagram for a 1050 steel containing 0.9% Mn. Note that above about 550°C, ferrite must form before any pearlite can. Data from *ASM Metals Handbook*, 8th ed., v. 8, ASM (1973).

ISOTHERMAL TRANSFORMATION

Figure 12.12 is the isothermal transformation diagram for a 1045 steel. It describes the transformation of austenite when austenite is suddenly cooled to a temperature below the equilibrium transformation temperature. It should be noted that for temperatures above about 600°C, ferrite forms before any pearlite can form. The rate of transformation depends on the rate of nucleation of pearlite colonies and the rate of their growth. In general, the nucleation rate increases with greater cooling below the equilibrium temperature of 727°C. Because growth depends on diffusion, the growth rate decreases at lower temperatures. As a result, the overall transformation rate first increases as the temperature is decreased below 727°C to about 600°C and then decreases below this.

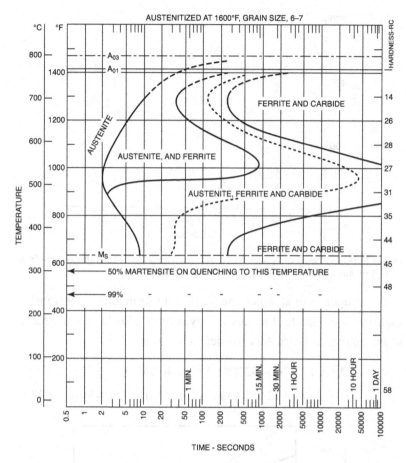

Figure 12.13. Isothermal transformation diagram for a 4340 steel containing 0.42% C, 0.78% Mn, 1.79% Ni, 0.80% Cr, and 0.33% Mo. From *The Making, Shaping and Treating of Steels*, 9th ed., ibid.

Chromium and molybdenum delay the transformation of austenite to pearlite more than to bainite, as shown in Figure 12.13 for a 4340 steel.

The temperature at which martensite starts to form, M_s, and at which the transformation is essentially complete, M_f, are lowered by increased carbon content, as shown in Figure 12.14. Increased alloy content has the same effect. The M_f temperature should be regarded

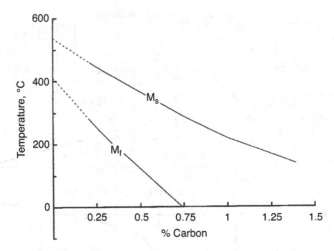

Figure 12.14. Increased carbon content lowers the M_s and M_f temperatures. Data from A. R. Troyano and A. B. Greninger, *Met. Prog.*, v. 50 (1946).

as the temperature at which the martensite formation is 99% rather than 100% complete. Increased alloy content also lowers the M_s as shown in Figure 12.15. An empirical relation is

$$M_s(°C) = 539 - 423(\%C) - 30.4(\%Mn)$$
$$- 12.1(\%Cr) - 17.7(\%Ni) - 7.5(\%Mo). \quad (12.1)$$

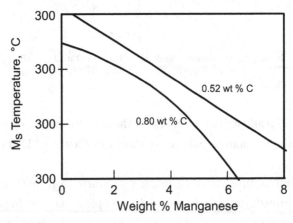

Figure 12.15. Increased alloy content lowers the M_s temperature. Data from J. V. Russell and F. T. McGuire, *Trans. ASM*, v. 33 (1944).

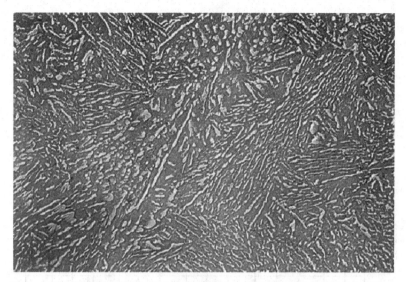

Figure 12.16. Electron micrograph of bainite formed at 700°F, 7500X. From *The Making, Shaping and Treating of Steels*, ibid.

BAINITE

If the γ → martensite transformation occurs below about 600°C, the resulting structure is called *bainite*. Like pearlite, it consists of ferrite and carbide, but the carbide is not in the form of lamellae. Instead, it consists of very fine isolated particles. Figure 12.16 is an electron photomicrograph of bainite. The size of the carbide particles is finer with lower temperatures of transformation.

CONTINUOUS COOLING DIAGRAMS

For most applications, transformation during continuous cooling is of greater direct interest than isothermal transformation. During continuous cooling, part of the time is spent at high temperatures at which the transformation rates are slow. Therefore, the time to transform is longer for continuous cooling than for isothermal treatment.

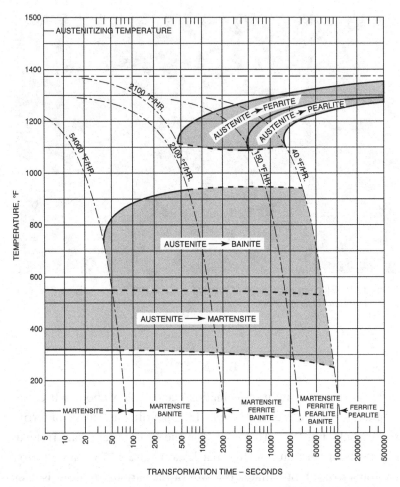

Figure 12.17. Continuous cooling transformation diagram for 4340 steel. From *The Making, Shaping and Treating of Steels*, ibid.

The lines representing the start of pearlite and bainite formation are shifted to longer times and lower temperatures. Figure 12.17 shows the continuous cooling transformation diagram for a 4340 steel. Note that because the alloying elements delay the pearlite formation more than the bainite formation, bainite can be formed during continuous cooling.

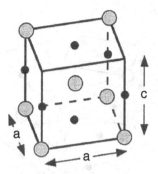

Figure 12.18. Martensite unit cell. The black dots indicate possible sites for carbon atoms.

MARTENSITE

Martensite has a body-centered tetragonal lattice (Figure 12.18). It can be thought of as supersaturated ferrite, the excess carbon causing the lattice to become elongated in one direction. Two unit cells of martensite form from one unit cell of austenite (Figure 12.19).

The amount of distortion of the lattice depends on the carbon content (Figure 12.20). The effect of carbon on the lattice parameter of martensite in nm is given by

$$c = 0.2861 + 0.0166x \qquad (12.2)$$

and

$$a = 0.2861 - 0.0013x, \qquad (12.3)$$

where x is the wt % carbon. The lattice parameter of austenite is

$$a = 0.3548 + 0.0044x. \qquad (12.4)$$

There is a relationship between the orientations of martensite and the austenite from which it formed. The Kurdjumov-Sachs relation, $\{101\}_M // \{111\}_A$ and $<111>_M // <110>_A$, with $\{225\}_A$ habit planes is valid for carbon contents of between 0.9 and 1.4%. The habit plane depends on the composition, as shown in Figure 12.21.

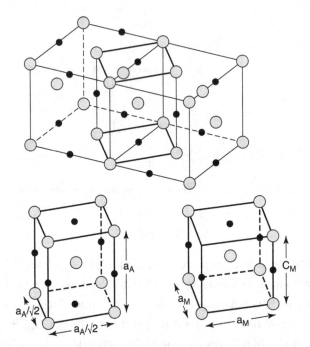

Figure 12.19. Relation between the martensite unit cells and the austenite from which it forms. Note that two martensite unit cells are formed for each austenite unit cell and that $c_M > a_A/\sqrt{2}$ and $c_M < a_A$. From W. F. Hosford, *Physical Metallurgy*, 2nd ed., CRC Press (2010).

Martensite is very hard, the hardness increasing with carbon content up to a value of about Vickers 850 (Rc65) at 0.8% carbon (Figure 12.22). The hardness of martensite is independent of the amount of alloying elements present.

RETAINED AUSTENITE

With the low M_f temperatures of higher-carbon steels, austenite is retained at room temperature, as shown in Figure 12.23.

Figure 12.24 shows that the hardness of quenched structures reaches a maximum amount at about 0.8% C and then decreases, even though the hardness of martensite does not drop. The decrease of

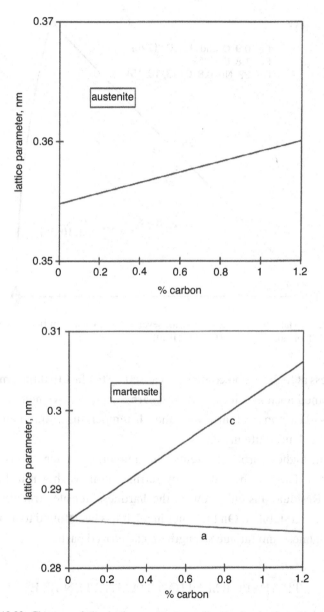

Figure 12.20. Changes of the lattice parameters of austenite (top) and martensite (bottom) with carbon content. Note that as the carbon content of martensite approaches zero, both c and a approach the lattice parameter of bcc ferrite. Data from C. S. Roberts, *Trans. AIME*, v. 197 (1953).

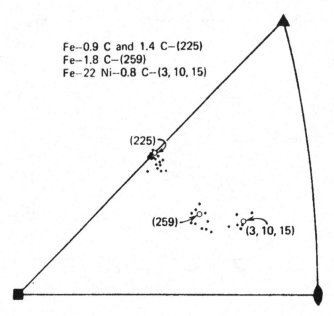

Figure 12.21. Habit plane of martensite for several steels. Data from B. A. Greninger and A. R. Troiano, *Trans. AIME* v. 140 (1940).

hardness at high carbon contents is caused by the fact that the amount of retained austenite increases with carbon content. Remember that increased carbon content lowers the M_s temperature, so the amount of retained austenite increases.

With higher carbon contents, the amount of retained austenite increases. This can be reduced by refrigeration, as shown in Figure 12.25. Residual austenite lowers the hardness, tensile strength, and dimensional stability. On the other hand, it has been found to increase the toughness and fatigue strength on carburized parts.

TRANSFORMATION TO MARTENSITE

Transformation of austenite to martensite occurs almost instantaneously. Martensite platelets grow at almost the speed of sound, taking only 10^{-7} to 10^{-5} seconds to form. The transformation involves

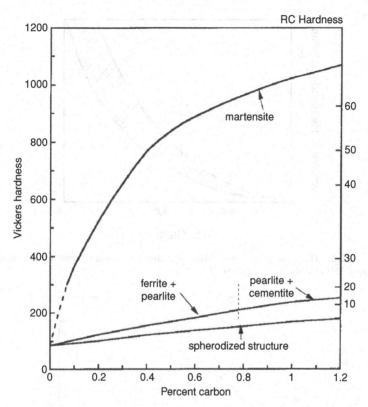

Figure 12.22. The hardness of martensite increases with carbon content up to a maximum of about Vickers 1000 at 0.8% C. From E. Bain and H. W. Paxton, ibid.

both shear and volume expansion. The elastic shear-strain energy is minimized by either slip or by twinning, as illustrated in Figure 12.26.

Martensite forms as discs or platelets (Figure 12.27). This decreases the volume of austenite that must accommodate the volume expansion and shear strain. The activation energy, ΔG, for martensite formation involves an increase of free energy caused by the new surface, $2\pi a^2 \gamma$, where a is the radius of the disc and γ is the interfacial austenite-martensite surface energy per area. There is also a shear-strain energy term, $(16\pi/3)(\gamma/2)^2 \mu a^2 c$, where μ is the shear modulus of the austenite and γ_s is the shear strain. The driving force for the

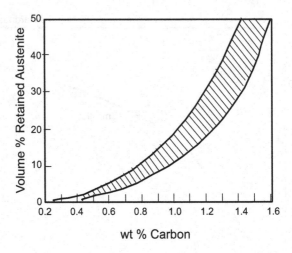

Figure 12.23. The amount of retained austenite after quenching increases with carbon content. Data from J. D. Verhoeven, ibid.

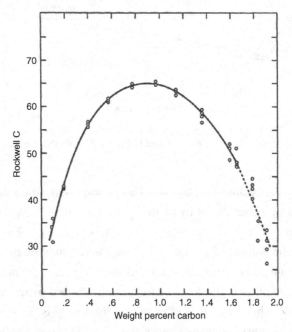

Figure 12.24. Above about 0.8% C, the hardnesses of quenched steels decrease with carbon content because of an increased amount of retained austenite. From A. Litwinchuk, F. X. Kayser, H. H. Baker, and A. Henkin, "The Rockwell C Hardness of Quenched High-Purity Iron-Carbon Alloys Containing 0.09 to 1.91% Carbon," *J. Mat. Sci.*, v. 11 (1976).

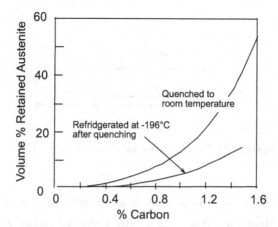

Figure 12.25. Refrigeration can be used to reduce the amount of retained austenite in plain carbon steels.

transformation is $(4/3)\pi a^2 c \Delta G_v$, where ΔG_v is the free energy difference per volume between the austenite and martensite. ΔG_v increases as the temperature decreases. The net value of ΔG is

$$\Delta G = 2\pi a^2 \gamma + 16\pi/3)(\gamma_s/2)^2 \mu a^2 c - (4/3)\pi a^2 c \Delta G_v. \quad (12.5)$$

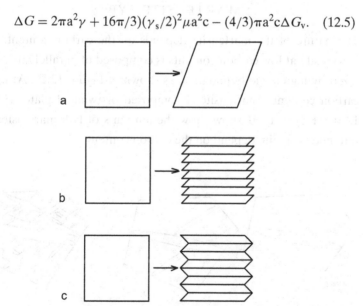

Figure 12.26. Most of the incompatibility of the large shear stain (a) can be accommodated by slip (b) or by twinning (c).

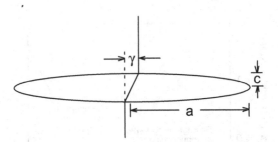

Figure 12.27. The dimensions and shear in Equation 12.5.

Each platelet that forms decreases the possible size of the next platelet, as shown in Figure 12.28. With a smaller radius, a, the next disc requires greater cooling below the M_s temperature to form. This decrease of the radius, a, of the next platelet requires greater cooling below the M_s temperatures.

MARTENSITE TYPES

The nature of the martensite depends on the carbon content. Lath martensite at low carbon contents is composed of parallel laths separated by high angle boundaries, as shown in Figure 12.29. At higher carbon contents, martensite is composed of twinned plates (Figure 12.30). Figure 12.31 shows how the amounts of lath martensite and retained austenite depend on the carbon content.

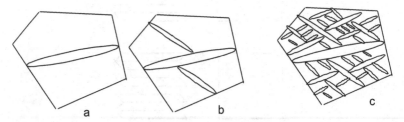

Figure 12.28. The first martensite platelet is limited in size by the austenite grain size (a). The next platelets to form must be smaller (b) and later platelets still smaller (c).

$100 \mu m$

Figure 12.29. Lath martensite. From G. R. Speich and W. C. Leslie, *Met. Trans*, v. 3 (1972).

Hypereutectoidsteels are normally austenitized just above the lower critical. In this way, the carbon content of the austenite is about 0.8% C.

The transition from lath to twinned plate martensite with carbon is illustrated in Figure 12.32.

SPECIAL HEAT TREATMENTS

When a steel is quenched, the outside cools much faster than the interior and therefore undergoes the martensitic transformation while the interior is still hot. When the interior later transforms to martensite, its volume expansion causes residual tensile stresses on the surface. These stresses can be avoided if the surface and interior transform to martensite at the same time. *Marquenching* (or *martempering*) is the

Figure 12.30. Twinned plate martensite. From G. R. Speich, and W. C. Leslie, ibid.

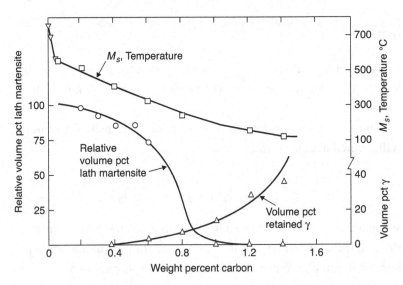

Figure 12.31. The amount of lath martensite decreases with high carbon content. The amount of retained austenite increases as the Ms temperature drops. Data for carbon steels. After G. R. Speich and W. C. Leslie, ibid.

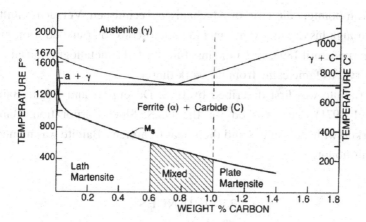

Figure 12.32. Transition from lath to twinned plate martensite. From A. R. Marder and G. Krauss, "The Morphology of Martensite in Iron-Carbon Alloys," *Trans. ASM*, v. 60 (1967), pp. 651–660.

name of a process in which steel parts are quenched into a molten salt bath at a temperature just above the M_s. After thermal equilibrium is reached, but before bainite starts to form, the parts are slowly cooled to form martensite. Because most of the part transforms to martensite at the same time, the residual stresses are greatly reduced.

Austempering is a process designed to produce bainite. Steel parts are quenched in a molten salt bath at a temperature just above the M_s and held at that temperature until transformation to bainite is complete.

MISCELLANY

Austenite is named for Sir William Roberts-Austen (1843–1902). (He was born William C. Roberts but added the name Austen after an uncle as a condition for inheritance.) Austen was an authority on metals and alloys. As such, he developed methods of analyzing alloys and recording temperatures.

Martensite was named for German metallurgist Adolf Martens (1850–1914). In 1873, he built his own microscope. In 1884, he was appointed director of Mechanisch-Technishen Versuchsantalt. Although his duties did not include microscopy, he pursued it on his own time until 1898, when the institute started a metallographic laboratory. His fame came from his work there.

Bainite was first described by E. S. Davenport and Edgar Bain (1891–1971), who worked for the U. S. Steel Corporation. Bain worked with steel alloys and their heat treatment. Bainite was named in his honor.

REFERENCES

ASM Metals Handbook, 8th ed., v. 8, ASM (1973).

E. G. Bain and H. W. Paxton, *Alloying Elements in Steel*, ASM (1939).

W. F. Hosford, *Physical Metallurgy*, 2nd ed., CRC Press (2010).

G. Krauss, *Steels: Heat Treatment and Processing Principles*, ASM (1990).

The Making, Shaping and Treating of Steels, 9th ed. U.S. Steel Corp. (1971).

D. A. Porter and K. E. Easterling, *Phase Transformations in Metals and Alloys*, 2nd ed., Chapman and Hall (1981).

A. K. Sinha, *Ferrous Physical Metallurgy*, Butterworth (1989).

J. D. Verhoeven, *Steel Metallurgy for the Nonmetallurgist*, ASM (2007).

R. Vilella, *Decomposition of Austenite by Diffusional Processes*, V. F. Zackay and I. Aaronson, eds. Interscience Publishers (1962).

13

HARDENABILITY

The influence of alloying elements on the rate of pearlite formation influences whether martensite will be formed when austenite is quenched because martensite can form only from austenite. If the formation of pearlite is delayed, more austenite will be available at the M_s temperature to transform to martensite. The term *hardenability* is used to describe this effect. We say that alloying elements increase the hardenability of steel, making it possible to harden them to greater depths.

JOMINY END-QUENCH TEST

Hardenability may be quantitatively described several ways. One of the simplest is the Jominy end-quench test in which a 4-in.-long, one-inch-diameter bar of the steel is austenitized and then placed in a fixture and cooled from one end with a specified water spray (Figure 13.1). The hardness is then measured as a function of distance from the quenched end. Figures 13.2 and 13.3 show the resulting curves for several steels.

Several features should be noted:

1. The cooling rate at the quenched end was fast enough to ensure 100% martensite in all the steels, so the hardness at the quenched end depends only on the carbon content.

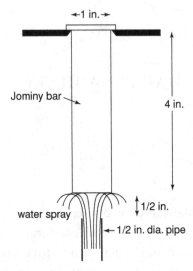

Figure 13.1. Jominy end-quench test. From W. F. Hosford, *Physical Metallurgy*, 2nd ed., CRC Press (2010).

2. The hardenability increases with the amount of alloying addition.
3. The hardenability increases with carbon content.

Although not indicated in these figures, a steel with a large austenite grain size has a higher hardenability than one with a fine austenite grain size. This is because with a larger austenite grain size, there are fewer nucleation sites for pearlite.

Jominy data can be used in several ways. For example the cooling rates at various positions in round bars during several types of quenches have been experimentally determined (Figure 13.4). For low-alloy steels, these cooling rates are independent of the steel composition. Therefore, if the Jominy curve for a steel is known, the hardness distribution in a quenched bar can be predicted. Consider a 2-in. diameter bar of 3140 steel quenched in "mildly agitated oil." The surface should cool at the same rate as a spot 5/16 in. from the end of a Jominy bar and therefore should have a hardness

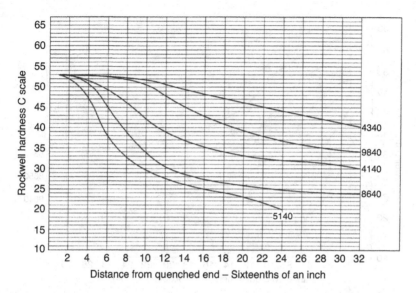

Figure 13.2. Hardnesses along Jominy bars of several steels. *The Making, Shaping and Treating of Steel*, 9th ed., United States Steel Corp. (1971).

	C	Mn	Ni	Cr	Mo
4340	0.40	0.70	1.65	0.70	0.20
9840	0.40				
4140	0.40	0.75	—	0.80	0.15
8640	0.40	0.75	0.40	0.40	0.15
5140	0.40	0.70	—	0.70	—

of Rc53. At the midradius, the equivalent Jominy distance is 10/16, so the hardness should be Rc43. At the center the radius, the equivalent Jominy distance is 11/16, so the hardness should be Rc42.

For a part having a complex shape, the hardness at various locations can be predicted by quenching the same shaped part made from known steel. Then by comparing the hardness at a given location with Jominy curve for this steel, the equivalent Jominy distance can be found, and this can be used to predict the hardness at this location of a similar part made from any steel.

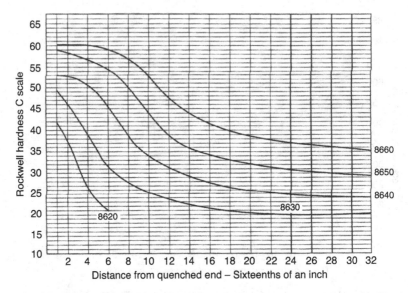

Figure 13.3. Hardness along Jominy bars of several steels of varying carbon contents. All contain 0.70 to 1.0% Mn, 0.4 to 0.7% Ni, 0.4 to 0.5% Cr, and 0.15 to 0.25% Mo. From *Making, Shaping and Treating of Steel*, 9th ed., U.S. Steel Corp. (1971).

For example, to predict the hardness of a 3-in.-diameter bar of 8660 steel, one can read from Figure 13.4 that in mildly agitated oil, the center of a 3-in.-diameter bar will cool at the same rate at 20/16 in. from the end of a Jominy bar. Figure 13.3 indicates that that it will harden to RC 39.

HARDENABILITY BANDS

The variation of compositions within a given grade of steel results in different Jominy curves. However, some steels are sold with the guarantee that the Jominy curves will lie within a certain band. Figure 13.5 is the hardenability band for an American Institute of Steel and Iron grade 4140 H steel.

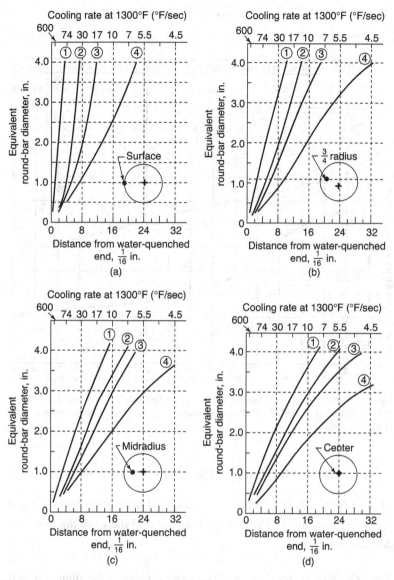

Figure 13.4. Cooling rates at different positions in round bars during several quenches related to positions along a Jominy bar.

1 = still water
2 = mildly agitated oil
3 = still oil
4 = mildly agitated molten salt

From R. A. Flinn and P. K Trojan, *Engineering Materials and Their Applications*, 4th ed., Houghton Mifflin (1990).

141

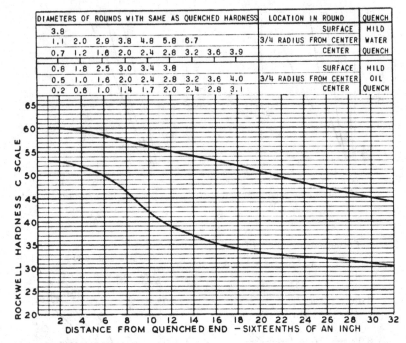

DIAMETERS OF ROUNDS WITH SAME AS QUENCHED HARDNESS									LOCATION IN ROUND	QUENCH
3.8									SURFACE	MILD
1.1	2.0	2.9	3.8	4.8	5.8	6.7			3/4 RADIUS FROM CENTER	WATER
0.7	1.2	1.6	2.0	2.4	2.8	3.2	3.6	3.9	CENTER	QUENCH
0.8	1.8	2.5	3.0	3.4	3.8				SURFACE	MILD
0.5	1.0	1.6	2.0	2.4	2.8	3.2	3.6	4.0	3/4 RADIUS FROM CENTER	OIL
0.2	0.6	1.0	1.4	1.7	2.0	2.4	2.8	3.1	CENTER	QUENCH

Figure 13.5. Hardenability band for an American Institute of Steel and Iron (AISI) 4140 H steel. From AISI Steel Products Manual, *Alloy Steel: Hot Rolled and Cold Finished Bars*, AISI (1970).

IDEAL DIAMETER CALCULATIONS

The hardenability of a steel can be predicted from its composition and grain size as follows: The *critical diameter* and the *ideal diameter* are defined as follows:

The *critical diameter*, D_c, for a steel and quench is the diameter that would harden to 50% martensite at center.

The *ideal diameter*, D_I, is the diameter that would harden to 50% martensite in an ideal quench.

An *ideal quench* is one for which there is no resistance to heat transfer from the bar to the quenching medium ($H = \infty$), so the surface comes immediately to the temperature of the bath.

Table 13.1. *Severity of quenches*

Agitation medium	Quench severity, H*		
	Oil	Water	Brine
None	0.25–0.30	0.9–1.0	2
Mild	0.30–0.35	1.0–1.1	2.0–2.2
Moderate	0.35–0.40	1.2–1.3	
Good	0.4–0.5	1.4–1.5	
Strong	0.50–0.80	1.6–2.0	
Violent	0.80–1.1	4.0	5.0

* H is defined as the heat transfer coefficient, h in Btu/(hr · ft^2°F) divided by the thermal conductivity of steel $= 20$ Btu/(hr · ft · °F) so its units are °F^{-1}.

The values of quench severity for various quenches are given in Table 13.1.

Grossman worked out a scheme for estimating the ideal diameter for a steel. Knowing the carbon content and grain size, one can find the base diameter for a plain-carbon steel (Figure 13.6). Each alloying

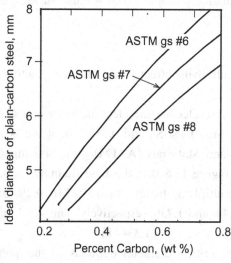

Figure 13.6. Dependence of the ideal diameter of a plain-carbon steel on its carbon content and austenite grain size. Adapted from M. A. Grossman, *Elements of Hardenability*, ASM (1952).

Table 13.2. *Multiplying factors*

Percent	Mn	Mo	Cr	Si	Ni
0.00	1.00	1.00	1.00	1.00	1.00
0.10	1.34	1.31	1.22	1.07	1.04
0.20	1.67	1.62	1.43	1.14	1.08
0.30	2.00	1.93	1.65	1.21	1.11
0.40	2.34	2.24	1.86	1.28	1.15
0.50	2.08	1.35	1.19	2.080	2.50
0.60	3.01		2.29	1.42	1.23
0.70	3.34		2.51	1.49	1.26
0.80	3.68		2.72	1.56	1.30
0.90	4.01		2.94	1.63	1.34
1.00	4.35		3.15	1.70	1.38
1.10	4.78		3.37	1.77	1.41
1.20	5.17		3.58	1.84	1.45
1.30	5.6			1.91	1.49
1.40	6.05			1.98	1.53
1.50	6.6			2.05	1.56
1.60	7.2			2.12	1.60

Data from *The Making, Shaping and Treating of Steel*, 9th ed., U.S. Steel Corp. (1971).

element has a multiplying effect. The multiplying factors are shown in Table 13.2.

For example, to calculate the ideal diameter of a steel containing 0.45% C, 0.5% Mn, 0.2% Si, and 0.35% Cr, if the American Society for Testing and Materials (ASTM) grain size number is 6, one can read from Figure 13.5 that the base diameter is 0.25 in. Table 13.2 gives the multiplying factors for 0.5% Mn, 0.2% Si, and 0.35% Cr of 2.667, 1.140, and 1.756, respectively. The ideal diameter $D_I = (0.25)(2.667)(1.140)(1.756) = 1.3347$ in.

However, the critical diameter depends on the quench severity. Figures 13.7 and 13.8 show the relation between the critical and ideal

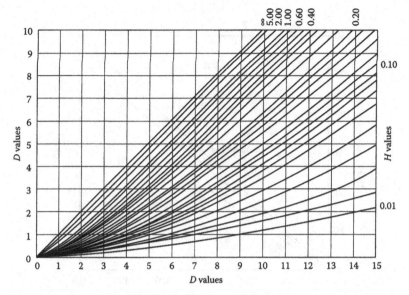

Figure 13.7. The relationship between the critical diameter and the ideal diameter for several severities of quenching. From *The Making, Shaping and Treating of Steels*, ibid.

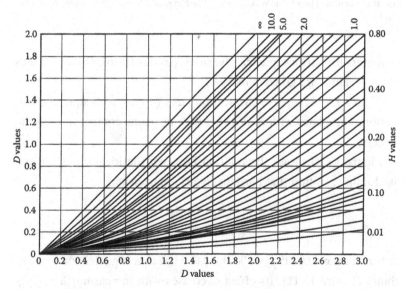

Figure 13.8. An enlargement of the low-diameter end of Figure 13.7. From *The Making, Shaping and Treating of Steel*, ibid.

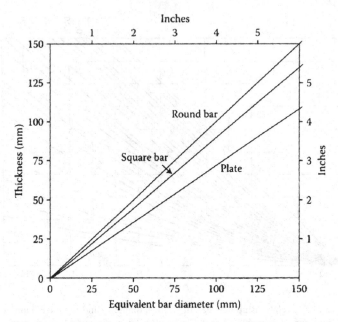

Figure 13.9. Comparison of sizes of square and round bars and plate cooling at the same rate at the center. Data from W. C. Leslie, *The Physical Metallurgy of Steels*, McGraw-Hill (1981).

diameters for various quench severities. (In this figure, *D value* means D_c, and values are given in inches.)

Figure 13.9 compares the section sizes of bars with square cross-sections and plates with round bars with the same cooling rates at the centers.

Figure 13.10 shows the relation between the ideal diameter and the Jominy position that hardens to 50% martensite.

BORON

Boron in range 0.0005 < wt % B <0.003 has a strong effect on hardenability (Figure 13.11). Its effect decreases with increasing % C (Figure 13.12) and depends on austenitizing temperature (Figure 13.13).

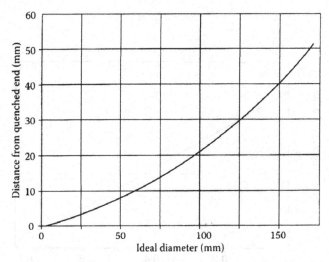

Figure 13.10. Relation between the ideal diameter and the Jominy position that hardens to 50% martensite. Data from D. J. Carney, "Another Look at Quenchants, Cooling Rates, and Hardenability," *Trans. ASM*, v. 46 (1954).

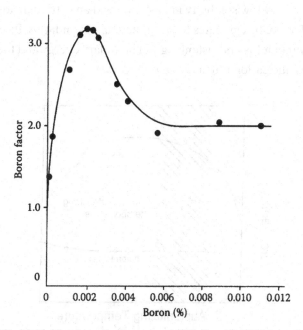

Figure 13.11. Multiplying factor for boron in a 0.20% C and 0.55% Mo steel. From G. F. Melloy, P. R. Simmon, and P. P. Podgurski, "Optimizing the Boron Effect," *Met. Trans.*, v. 4 (1973).

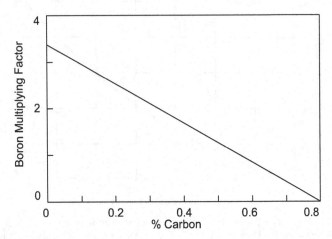

Figure 13.12. Effect of carbon on the multiplying factor for boron in 0.8% Mn steels. Data from W. T. Cook, *Met. Tech.*, v. 1 (1974).

Boron has very low solubility in austenite and an extremely low solubility in ferrite. It segregates to austenite grain boundaries. Its effect is probably that it lowers austenite grain boundary energy and therefore decreases nucleation rate at grain boundaries.

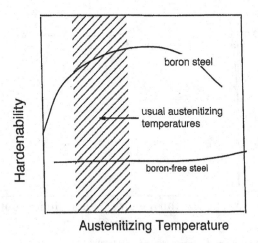

Figure 13.13. The effect of boron decreases for high austenitizing temperatures. Adapted from R. A. Grange and J. B. Mitchell, "On the Hardenability Effect of Boron in Steel," *Trans. ASM*, pp. 157–185.

MISCELLANY

From 1932 to 1936, E. C. Bain and coworkers did pioneering work on the kinetics of decomposition of austenite and presented their results in the form of TTT curves. In 1934, B. F. Shepherd proposed that the hardenability of a steel could be assessed by quenching a bar of steel and measuring the depth of hardening. Walter Jominy and A. L. Boegehold first published the details of the end-quench test in 1938. About the same time, M. A. Grossman and coworkers proposed the concept of an ideal diameter. Robert Mehl explained the physics behind the affects of alloys on hardenability in 1939.

REFERENCES

M. A. Grossman, *Elements of Hardenability*, ASM (1952).
G. Krauss, *Steel – Heat Treatment and Processing Principles*, ASM (1990).
W. C. Leslie, *The Physical Metallurgy of Steel*, McGraw-Hill (1981).
The Making, Shaping and Treating of Steel, U.S. Steel Corp. (1971).
A. K. Sinha, *Ferrous Physical Metallurgy*, Butterworths (1989).

14

TEMPERING AND SURFACE HARDENING

TEMPERING

Martensite is brittle. To make it tougher, it is normally heated to *temper* it. Tempering involves a complex series of reactions that gradually break down martensite. What happens is usually described in stages. The first stage occurs at the lowest temperature (shortest time) and involves transformation of retained austenite. In the second stage, carbon is redistributed within the martensite to dislocations. Generally, stress relief occurs during this stage. Precipitation of ε carbide ($Fe_{2.4}C$) and η carbide (Fe_2C) from the martensite comprises the third stage. This precipitation lowers the carbon content of the martensite. In stage four, remaining austenite decomposes to cementite (Fe_3C) and ferrite. Finally, in stage five, the transition carbides and low-carbon martensite form more ferrite and cementite. These reactions overlap.

There is a gradual loss of hardness throughout tempering (except stage one) at increasing temperatures. This is shown as a function of the carbon content, as shown in Figure 14.1. Figure 14.2 shows the effect of tempering temperature on the properties of 4340 steel. The amount of tempering depends on the carbon content (Figure 14.3) as well as time and temperature (Figure 13.4), although the effect of time is much less than that of temperature.

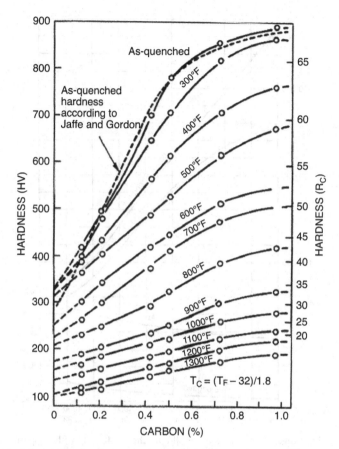

Figure 14.1. Tempering martensite in 4340 steel. The hardness drops with increasing temperature. One half hour tempering time. From R. A. Grange, C. R. Hribal, and L. F. Porter, "Hardness of Tempered Martensite in 12 Carbon and Low-Alloy Steels," *Met. Trans. A*, v. 8A (1977), pp. 1775–1785.

The extent of tempering can be treated by an Arrhenius equation. The time, t, to reach a given hardness is

$$t = A \exp(+Q/RT), \qquad (14.1)$$

or the hardness, H, can be expressed as

$$H = f[t \exp(-Q/RT)]. \qquad (14.2)$$

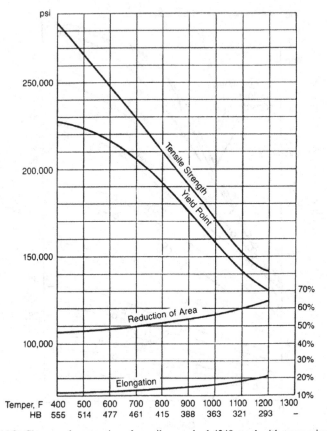

Figure 14.2. Change of properties of an oil-quenched 4340 steel with tempering temperature. From *Modern Steels and Their Properties*, Bethlehem Steel Corp. (1972).

The activation energy for tempering of the steel in Figure 14.3 can be found by comparing the time-temperature combinations that result in a hardness of RC60. The Arrheneous equation $t = A \exp(Q/RT)$, $t_2/t_1 = \exp[(Q/R)(1/T_2 - 1/T_1)])$, so $Q = R \ln(t_2/t_1)/(1/T_2 - 1/T_1)$. Taking points at RC60 in Figure 14.2, $t_2 = 800$ min @ $T_2 = 250°C = 523$ K and $t_1 = 0.6$ min @ $T_2 = 315°C = 588$ K. $Q = 7.134 \ln(800/0.6)/(1/523-1/588) = 249$ kJ/mole.

Substituting $T_3 = 220°C = 493$ K, into $t_3/t_2 = \exp[(Q/R)(1/T_3 - 1/T_2)] = 35.2$ min.

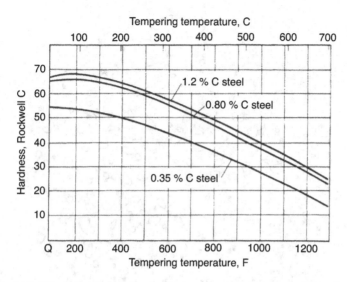

Figure 14.3. Effect of carbon content and tempering temperature on the hardness of plain-carbon steels. From A. K. Sinha, *Ferrous Physical Metallurgy,* Buttersworth 1989.

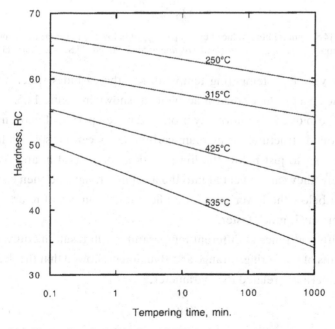

Figure 14.4. Tempering of a 0.82% C and 0.75% Mn steel. Adapted from data of E. C. Bain, *Functions of Alloying Elements in Steel,* ASM (1969).

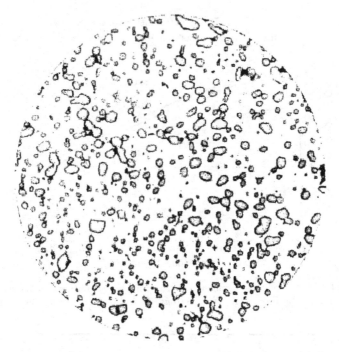

Figure 14.5. Spheroidite produced by tempering just below the lower critical temperature. From *The Making, Shaping and Treating of Steel*, 9th ed., U.S. Steel Corp. (1971).

At very high tempering temperatures, the carbides spheroidize,
producing a product called spheroidite as shown in Figure 14.5.

Spheroidite is not normally produced by tempering of martensite.
Hot-rolled structures of medium-carbon steels can be spheroidized
by heating to just below the lower critical for several hours. Alternatively, they can be heated into the austenite range and then cooled
to just below the lower critical and held for a number of hours. The
former way is much faster.[*]

Different times at different temperatures will result in equivalent
amounts of tempering. Grange and Baughman showed that the degree
of tempering is related to a parameter,

$$F = T(18 + \log_{10} t), \tag{14.3}$$

---

[*] J. O'Brien and W. Hosford, "Spheroidizing Cycles for Medium Carbons Steels,"
Metallurgical and Materials Transactions, v. 33 (2002).

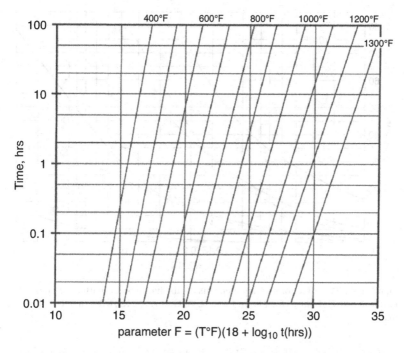

Figure 14.6. Time-temperature combinations that result in the same parameter describing the extent of tempering. Data from R. A. Grange and R. W. Baughman, "Hardness of Tempered Martensite in 12 Carbon and Low-Alloy Steels," *Trans. ASM*, v. 48 (1956), pp. 165–197.

where T is the temperature in °F plus 460, and t is the time in hours. Figure 14.6 shows this relation.

Alloying elements slow the rate of tempering and reduce the decrease in hardness. Figure 14.7 illustrates this. This slowing of tempering by alloying elements is partially a result of the fact that substitutional elements diffuse more slowly than carbon. Many of the alloying elements also form harder carbides than iron.

It is interesting to compare the tempering of steel with precipitation hardening of nonferrous alloys. Both processes involve the precipitation of fine particles from a supersaturated solid solution (martensite can be thought of as a supersaturated solution of carbon in ferrite). In tempering, the precipitates cause hardening, but the loss of solid solution has a greater effect, and so the net result is a

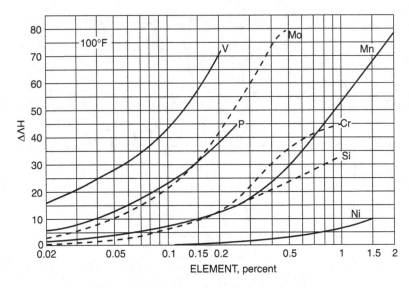

Figure 14.7. The effect of alloying elements on the hardness of martensite after tempering at 538°C (1000°F) for 1 hour. From R. A. Grange, C. R. Hribal, and L. F. Porter, ibid.

softening. In the case of precipitation hardening of nonferrous metals, the increase of hardness from the precipitates more than makes up for the loss of solid-solution strengthening. In the case of tempering, the loss of solution hardening is much greater than the precipitation-hardening effect.

In alloy steels, some of the retained austenite may transform to martensite on cooling from tempering. This untempered martensite will lower the toughness. To alleviate this problem, steels sometimes are double tempered. The second treatment tempers the martensite formed on cooling from the first tempering.

SECONDARY HARDENING

There is a strong effect of alloying elements on how much softening occurs. With large amounts of Cr, Mo, and W, there may be secondary hardening. Steels containing these alloys are resistant to tempering.

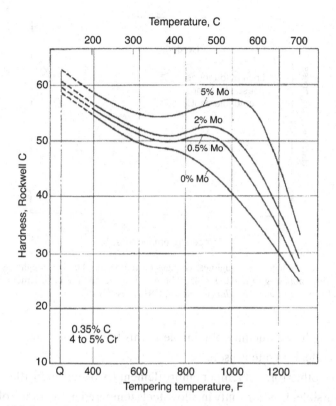

Figure 14.8. Increasing molybdenum content results in secondary hardening during tempering. From *The Making, Shaping and Treating of Steel*, ibid.

When carbides do precipitate at high temperatures, they have a strong precipitation-hardening effect and are resistant to coarsening. Figure 14.8 shows the effect of molybdenum. This effect is important in high-speed tool steels.

TEMPER EMBRITTLEMENT

There are two forms of temper embrittlement. One is the 500°F (350°C) embrittlement that occurs in low-alloy steels after tempering in the 250 to 450°C temperature range. See Figure 14.9. The fractures are intergranular. This is particularly troublesome if the combination

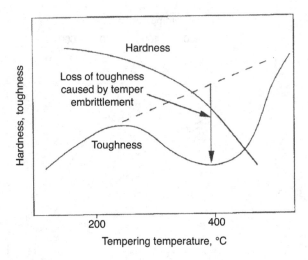

Figure 14.9. Hardness and toughness changes on tempering. The low toughness associated with tempering in the 250 to 450°C range is called 500°F embrittlement. From
W. F. Hosford, *Physical Metallurgy*, 2nd ed., CRC Press (2010).

of strength and ductility obtainable in this temperature range is necessary. Here bainite is useful.

The other form of temper embrittlement (two-step embrittlement)
is reversible. It occurs only in alloy steels tempered in the range of 600
to 700°C and slowly cooled through 600 to 350°C. It is caused by Sb
and P (and secondarily by Sn and As). Ni, Mn, Cr, and Si aggravate
the embrittlement. Mo, Ti, and Zr delay its onset. Heating to more
than 600°C followed by a rapid cooling reverses the embrittlement.

CARBURIZING

There are several techniques for hardening the surfaces of steels.
These include diffusing carbon (carburizing) and nitrogen (nitriding)
into the surface. Surfaces can also be hardened by heat treating the
surface without heat treating the core (this is called case hardening).

In carburizing, the steel is heated into the austenite region (900–
930°C) and subjected to a carburizing atmosphere. This is usually a
mixture of carbon monoxide and water vapor. Carbon is deposited in

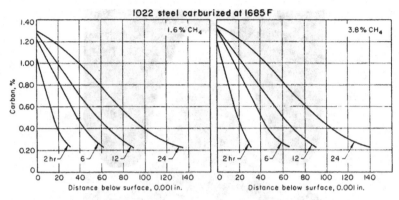

Figure 14.10. Depth of carburizing for several times. Note that the depth for a fixed concentration is proportional to √(time). From *ASM Metals Handbook*, 8th ed., v. 2 ASM (1964).

the steel by the reaction $CO + H_2 \rightarrow C + H_2O$. The CO/H_2 ratio is controlled to give carbon potential in equilibrium with 0.9% C in the austenite. This is often a two-step process, which initially uses higher C potential than is desired in the final product, followed by a period with a lower carbon potential, which allows time for diffusion. If % C is too high, hardness falls because of retained austenite, as shown in Figure 13.12.

In steels with a high carbon content, martensite formation causes a large volume expansion, but the volume expansion accompanying bainite formation is small. Carburizing leads to compressive residual stresses because the lower % C in the core allows it to transform first, so the martensite formation in the case will cause residual stresses.

CARBURIZING KINETICS

Carburizing is usually done at temperatures above 910°C. When the surface concentration of a steel containing C_o is suddenly increased to C_s and held at that level for a time, t, the concentration, C, at a distance, x, from the surface is given by

$$C = C_s - (C_s - C_o)\text{erf}[x/2\sqrt{(Dt)}]. \qquad (14.4)$$

This is illustrated in Figure 14.10.

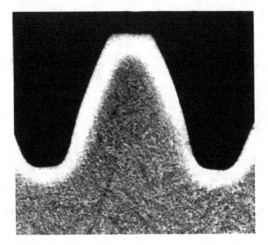

Figure 14.11. Cross-section of a gear of a steel containing 0.016% C, 0.6% Mn, 1.65% Cr, and 3.65% Ni carburized at 920°C, and austenitized at 830°C, and quenched and tempered at 150°C. The case depth is 0.8 to 1.0 mm. From *ASM Metals Handbook*, 9th ed., v. 9 (1985).

The experimental data in Figure 14.10 show that the depth of hardening is proportional to $\sqrt{t}$. Figure 14.11 is the cross-section of a carburized gear.

The situation would be very different if carburization were attempted at temperatures below 910°C. A schematic phase diagram and concentration profile are shown in Figure 14.12. The concentration gradient in the austenite is $(C_s - C_\gamma)/x$, where C_s and C_γ are the carbon concentrations at the surface and that in equilibrium with ferrite. The flux of carbon is

$$J = -D_\gamma \, dC/dx = -D_\gamma(C_s - C_\gamma)/x. \qquad (14.5)$$

This flux represents the accumulation of carbon associated with the advance of the carburized layer,

$$J = (C_\gamma - C_0) \, dx/dt, \qquad (14.6)$$

where C_0 is the composition of the steel. Equating and rearranging,

$$x \, dx = D_\gamma[(C_\gamma - C_0)/(C_s - C_\gamma)] \, dt.$$

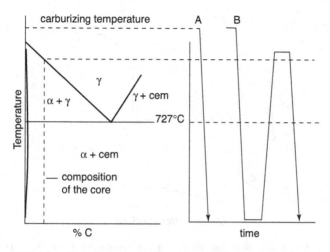

Figure 14.12. The effect of heat treatment on the core and case of a carburized part.

Integrating,

$$x = \{2[D_\gamma[(C_\gamma - C_o)/(C_s - C_\gamma)]t\}^{1/2}. \qquad (14.7)$$

The depths of hardening typically vary from 0.020 to 0.060 in. (0.5 to 1.5 mm). The depth is controlled by the diffusion of carbon into the surface. All solutions to Fick's second law predict that concentration is a function of $x/\sqrt{(Dt)}$, where x is the distance at which a certain concentration occurs, D is the diffusivity, and t is the time. Hardness is a function of carbon concentration, so the depth, x, at which a given hardness will be found depends only on D and t.

For a fixed hardness,

$$x = A\sqrt{(Dt)}, \qquad (14.8)$$

where A is a constant. To double the depth of hardening at fixed carburizing conditions, the time must be increased fourfold.

The high temperatures required for carburizing cause growth of the austenite grain size in the core. This can be alleviated with a second austenitization at a lower temperature, as illustrated in Figure 14.12. If the carbon content of the surface becomes too high, it may not completely transform to martensite.

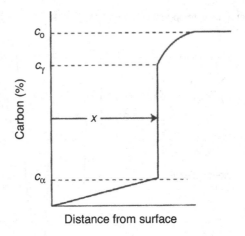

Figure 14.13. Concentration gradient during decarburization. From W. F. Hosford, ibid.

KINETICS OF DECARBURIZATION

When steel is heated in air, the surface loses carbon. Consider the decarburization at a temperature below 910°C of a steel having a carbon content of Co when it is heated into the austenite (γ) region and held in air. At this temperature, the reaction $2C + O_2 \rightarrow 2CO$ effectively reduces the carbon concentration at the surface to zero. A layer of α forms at the surface and into the steel to a depth, x. The concentration profile near the surface is shown in Figure 14.13. The concentration gradient is $dC/dx = -C_\alpha/x$, where C_α is the carbon content of the α in equilibrium with the γ. Fick's first law gives the flux,

$$J = -D_\alpha dc/dx = D_\alpha C_\alpha/x. \qquad (14.9)$$

Alternatives to carburizing in a mixed CO/H_2 atmosphere include mixed CO/CO_2, low-pressure methane atmosphere, and pack carburizing. Pack carburizing involves heating the part to be carburized while it is packed in coke or charcoal. In pack carburizing, almost all of the transport of carbon from the coke to the steel is by CO.

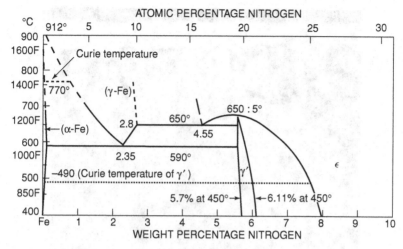

Figure 14.14. The iron-nitrogen phase diagram. From *ASM Metals Handbook*, 8th ed., v. 8, ASM (1973).

CARBOAUSTEMPERING

Carboaustempering is a relatively new process in which steel is austempered after carburization. Because the core has a low carbon content, its Ms temperature is much higher than that of the case, so it will transform to martensite during an interrupted quench to a temperature above the Ms of the case. Then holding will allow the case to transform to bainite.

NITRIDING

In nitriding, nitrogen is diffused into steel at temperatures below the lower critical (500–570°C). The sources of nitrogen include NH_3, NH_3-N_2 mixtures, NH_3 mixed with endothermic gas (40% N_2, 20% CO, 40% H_2), and NH_3-N_2-CO_2 mixtures. Diatomic nitrogen, N_2, will not dissolve in steel. Figure 14.14 shows the Fe-N phase diagram. The depths of nitrided layers are less than the depths of carburized layers. A depth of 0.01 to 0.015 in. (0.25–0.37 mm) can be obtained in 48

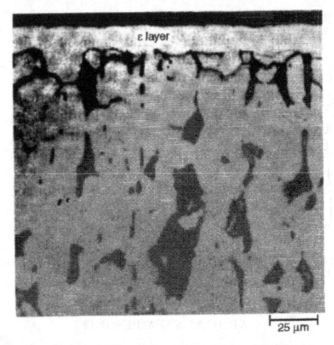

Figure 14.15. Surface of a 1018 steel after carbonitriding at 570°C for 3 hours. Note how thin the nitrided layer is. From *ASM Metals Handbook*, v. 4, 9th ed. ASM (1995).

hours. Figure 14.15 is a microstructure of a nitrided steel. In unalloyed steels, the hardness is a result of ε-carbide (Fe_3N). Steels used for nitriding usually contain aluminum and chromium or molybdenum. A typical nitriding composition is 1.0% Al, 0.25% C, 0.5% Mn, 0.25% Si, 1.0% Cr, and 0.2% Mo.

CARBONITRIDING

This is done in molten cyanide salts or in a NH_3-containing gas, which allow diffusion of both carbon and nitrogen into the steel. The temperature is below 590°C. The nitrogen acts mainly to form Fe_4N (ε-phase), which is responsible for the hardness.

CASE HARDENING WITHOUT
COMPOSITION CHANGE

If a strong heat source is applied to a part, the surface can be heated into the austenite temperature range while the interior remains cold. As soon as the heat source is removed, the unheated interior provides the quench necessary for martensite formation. The key is a rapid heating of the surface. With high-frequency induction, heating is restricted to the surface. Surface heating may also be accomplished with a laser or an intense flame.

FURNACE ATMOSPHERES

Different gases have different characteristics. Oxygen is oxidizing. It reacts with dissolved carbon to form carbon monoxide, $O_2 + C \rightarrow$ CO. It must be excluded for bright anneals. CO_2 is also oxidizing and decarburizing.

CO is a source and carrier of carbon. It is reducing. The CO_2/CO ratio controls whether the atmosphere is oxidizing or reducing. A ratio of 0.6 is oxidizing at 800°C; a ratio of 0.4 is not oxidizing but will decarburize a 1% C steel. For low-carbon steels, a ratio of 0.5 is used for bright anneals. Lower ratios may carburize.

Nitrogen is inert. Hydrogen, H_2, is highly reducing: Water, H_2O, is oxidizing. Hydrocarbons and proprietary mixtures are sources of carbon. These include methane (*endothermic gas*), which is a mixture of 20% CO, 40% H_2O, 40% N_2 plus methane and is used for carburizing. Ammonia and NH_3, when decomposed in an arc, are used as a source of nitrogen for nitriding. A mixture of ammonia and endothermic gas can be used for carbonitriding.

REFERENCES

ASM Metals Handbook, 8th ed., v. 8, ASM (1973).
ASM Metals Handbook, 9th ed., v. 9, ASM (1985).

ASM Metals Handbook, 9th ed., v. 4, ASM (1995).

ASM Metals Handbook, 8th ed., v. 2 ASM (1964).

E. C. Bain, *Functions of Alloying Elements in Steel*, ASM (1969).

The Making, Shaping and Treating of Steel, 9th ed., U.S. Steel Corp. (1971).

W. F. Hosford, *Physical Metallurgy*, 2nd ed., CRC Press (2010).

A. K. Sinha, *Ferrous Physical Metallurgy*, Butterworth (1989).

G. Krauss, *Steels: Heat Treatment and Processing Principles,* ASM (1990).

J. D. Verhoeven, *Steel Metallurgy for the Nonmetallurgist*, ASM (2007).

15

LOW-CARBON SHEET STEEL

SHEET STEELS

Low-carbon sheet steel may be finished by hot rolling or cold rolling. *Hot-rolled* steel has a rougher surface finish that limits its use to applications in which surface appearance is not important (e.g., auto underbodies and firewalls). *Cold-rolled* steels are almost always recrystallized before sale to fabricators. They are therefore softer than hot-rolled steels and have a much better surface finish.

Usually aluminum is added to molten low-carbon steel as it is poured. Without the addition of aluminum, dissolved oxygen would react with dissolved carbon to from CO. This reaction is violent, the CO bubbles causing steel droplets to fly into the air where they ignite. This process is called *killing* the steel and the steel called *killing*. AKDQ (aluminum-killed, drawing quality) is the designation for most of the steel sheet used in forming operations. The amount of oxygen that can dissolve in molten steel decreases with increasing carbon contents, as shown in Figure 15.1. Therefore, killing is not required for higher carbon contents. Today almost all low-carbon steel is continuously cast. AKDQ sheets usually contain 0.03% C or less.

The aluminum that does not react to form Al_2O_3 can react after solidification with dissolved nitrogen to form AlN.

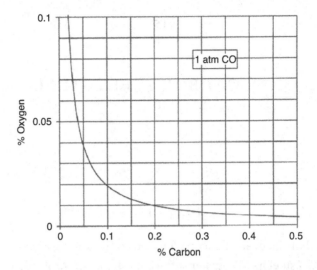

Figure 15.1. Equilibrium between oxygen and carbon in molten steel at 1540°C. With high carbon contents, the oxygen solubility is low and vice versa.

STRENGTH

Higher carbon and alloy content promote higher yield strength (YS) but lower strain-hardening exponent n (Figure 15.2). The R-value is affected by the rolling and annealing cycles. With growing interest in reducing the weight of automobiles, the use of higher-yield strength sheet has increased.

Properties of sheet steels vary from the interstitial-free (IF) steels with tensile strengths (TS) of 145 MPa and tensile elongation (EL) of 50% to martensitic steels with tensile strengths of 1400 MPa and tensile elongations of 5%. Table 15.1 lists typical properties of various grades of sheet steels. In general, the tensile elongation decreases with increased yield strength as shown in Figure 15.3.

GRADES OF LOW-CARBON STEEL

Among the grades of low-carbon steels are the following.

Interstitial-free (IF) steels: these are steels from which carbon and nitrogen have been reduced to extremely low levels (less than

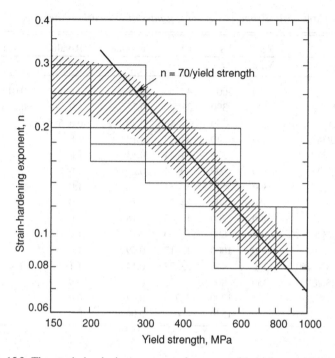

Figure 15.2. The strain-hardening exponent decreases with higher yield strengths. Adapted from S. P. Keeler and W. C. Brazier in *Micro Alloying 75*, Union Carbide (1977).

0.005%). After vacuum degassing, titanium is added to react with any carbon or nitrogen in solution. Titanium reacts preferentially with sulfur so the stoichiometric amount of titanium that must be added to eliminate carbon and nitrogen is

$$\%Ti = (48/14)(\%N) + (48/32)(\%S) + (48/12)(\%C). \quad (15.1)$$

A typical composition is 0.002% C, 0.0025% N, 0.025% Ti, 0.15% Mn, 0.01% Si, 0.01% P, 0.04% Al, and 0.016% Nb. Niobium may be used instead of titanium, but it is more expensive. IF steels typically have very low strengths. The advantage of these steels is that they are formable. Control of crystallographic texture is also fundamental in producing exceptional deep drawability. Typically $\bar{R} = 2.0$ (*R*-values of traditional aluminum-killed steels rarely exceed 1.8.) Formability is also enhanced by high *n* values (≈ 2.5)

Table 15.1. *Properties of some grades of low-carbon sheet steels*

Steel	YS	TS	El	n-value	R-value	m-value
MPa	MPa	%				
IF	150	300	45	0.28	2+	0.015
IF w/P	220	390	37	0.21	1.9	0.015
AKDQ	180	350	32–40	0.20–0.22	0.015	
BH210/340	210	340	34–39	0.18	1.8	
BH260/370	260	370	29–34	0.13	1.6	
DP280/600	280	600	30–34	0.21	1.0	
DP300/500	300	500	30–34	0.16	1.0	
DP350/600	350	600	24–30	0.14	1.0	
DP400/700	400	700	19–25	0.14	1.0	
DP500/800	500	800	14–20	0.14	1.0	
DP700/1000	700	1000	12–17	0.09	1.0	
HSLA350/450	350	450	23–27	0.14	1.1	0.005–0.01
TRIP450/800	450	800	26–32	0.24	0.9	
Mart950/1200	950	1200	5–7	0.07	0.9	
Mart1250/1520	1250	1520	4–6	0.065	0.9	

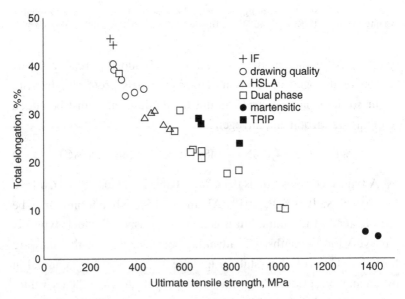

Figure 15.3. Decrease of total elongation with increasing strength.

Fine-grain sizes and higher strengths can be achieved by alloying with Nb. High-strength IF steels are solution hardened with small amounts of Mn, Si, and P. The tensile strength is increased 4 MPa by 0.1% Mn, 10 MPa by 0.1% Si, and 100 MPa by 0.1% P in solution. The presence of titanium reduces the phosphorus in solution by forming FeTiP. The increased strength comes at the expense of a somewhat reduced formability. A composition of 0.003% C, 0.003% N, 0.35% Mn, 0.05% P, 0.03% Al, 0.035% Nb, 0.2% Ti, and 0.001% B has the following properties: YS = 220 MPa, TS = 390 MPa, elongation = 37%, $\overline{R}$ = 1.9, and n = 0.21.

AK steels: dissolved carbon and oxygen in steels having carbon contents of 0.05 to 0.10% will react to form CO on freezing. This causes a violent rimming action as the CO bubbles are emitted, causing tiny drops of iron to burn in the air. Aluminum is added in the ladle to react with oxygen, removing it in the form of Al_2O_3, which rises to the surface and is scraped off. This "kills" the rimming action. Aluminum also ties up dissolved nitrogen as AlN. Because of the removal of nitrogen, strain aging only occurs at elevated temperatures. Typically, an aluminum-killed steel will have an n-value of about 0.22 and an R-value of 1.8.

Bake-hardenable steels: steels having enough carbon and/or nitrogen in solution to strain age at the temperatures used for paint baking are called *bake hardenable*. The low yield strength without a yield point before forming and the high strength caused by strain aging is useful for producing dent resistance or downsizing the thickness for weight reduction. Bake-hardenable steels are used in hoods, quarter-deck panels, roofs, doors, and fenders. With a low carbon level, they have good weldability.

High-strength low-alloy (HSLA) steels: HSLA steels are much stronger than plain-carbon steels. They are used in cars, trucks, cranes, bridges, and other structures for which stresses may be high. A typical HSLA steel may contain 0.15% C, 1.65% Mn, and low levels (under 0.035%) of P and S. They may also contain small amounts of Cu,

Ni, Nb, N, V, Cr, Mo, and Si. The term *micro-alloying* is often used because of the small amounts of alloying elements. As little as 0.10% niobium and vanadium can have profound effects on the mechanical properties of a 0.1% C, 1.3% Mn steel. The Mn provides a good deal of solid solution strengthening. The other elements form a fine dispersion of precipitated carbides in an almost pure ferrite matrix. Rapid cooling produces a fine grain size, which also contributes to the strength. Yield strengths are typically between 250 and 590 MPa (35,000–85,000 psi). The ductility and n-values are lower than in plain-carbon steels. HSLA steels are also more rust resistant than most carbon steels because of their lack of pearlite. An alloy with small amounts of Cu, called Corten, forms an adherent rust that is used architecturally.

Dual-phase steels: these steels have been heat treated to form islands of 5 to 15% martensite in a ferrite matrix. They replace many HSLA grades, having a lower initial yield strength but a higher rate of strain hardening. This makes them easier to form. Uses include front and rear rails, bumpers, and panels designed for energy absorption, for example.

TRIP (transformation-induced plasticity) *steels*: the microstructure of TRIP steels consists mainly of ferrite, but martensite, bainite, and retained austenite are also present. The various levels of these phases give TRIP steels their unique balance of properties.

During forming, the retained austenite transforms to martensite. This results in a high rate of work hardening that persists to higher strains, in contrast to that of dual-phase steel, which decreases at high strains. This causes enhanced formability. The carbon content controls the strain level of retained austenite-to-martensite transformation. With low carbon levels, transformation starts at the beginning of forming, leading to excellent formability and strain distribution at the strength levels produced. With high carbon levels, retained austenite is more stable and persists into the final part. The transformation occurs at strain levels beyond those produced during stamping and forming. Transformation to martensite occurs during subsequent deformation, such as a crash event, and provides greater crash energy absorption. Spot welding of TRIP steels is made more difficult by the alloying elements.

Complex-phase steels: these steels have a fine microstructure of ferrite with martensite and bainite. They are further strengthened by precipitation of niobium, titanium, or vanadium carbonitrides. They are used for bumpers and B-pillar reinforcements because of their ability to absorb energy.

Martensitic grades: the microstructures of martensitic grades are completely martensite. Tensile strengths vary between 900 and 1,500 MPa (130 and 220 KSI). These grades can be made directly at the steel mill by quenching after annealing or by heat treating after forming. Mill-produced material has low ductility and thus is typically is roll formed.

The carbon content controls the strength level. The tensile strength in MPa is approximately

$$TS = 900 + 2800(\%C). \qquad (15.2)$$

Manganese, silicon, chromium, molybdenum, boron, vanadium, and nickel are used in various combinations to increase hardenability. Typical applications for martensitic steels usually are those requiring high strength and good fatigue resistance, with relatively simple cross-sections, including door intrusion beams, bumper reinforcement beams, side sill reinforcements, and belt-line reinforcements.

Typical properties of low-carbon sheet steels are listed in Table 15.1.

Until recently, automobile bodies were made almost entirely from aluminum-killed low-carbon steels. The current emphasis on increased fuel economy accomplished through lighter vehicles has led to greatly increased use of the thinner gauges permitted by higher-strength steels. Low-carbon steels are rapidly becoming a minor part of auto bodies. The term *advance high-strength steels* (AHSS) has been applied to steels with yield strengths greater than 200 MPa.

The lower ductility of higher strength steel (Figure 15.3) results in lower formability. Much current research is aimed at increasing the ductility of higher-strength steels. One approach is to incorporate more austenite into the microstructures. Increased austenite levels together with more martensite and still finer grain sizes also raise the strength level.

WEATHERING STEEL

The best-known weathering steel is Corten, a proprietary product of U.S. Steel Corporation. There are two grades, differing slightly in composition. One contains 0.12% C and the other 0.16% C. Both contain copper (0.25–0.55%) and nickel (0.40–0.65%). These steels eliminate the need to paint in outdoor applications by forming a stable form of rust.

HEATING DURING DEFORMATION

Considerable heating can occur during the high strain rates necessary to form high-strength steels. Typical strain rates in automotive stampings is 10/s so little heat is transferred from the deforming steel. According to Equation 5.24, with a flow stress of 720 MPa and a strain of 0.50, the expected temperature rise is about 100°C. This temperature rise can lower the flow stress, appreciably as shown in Figure 15.4 and lead to necking at greatly reduced strains. There is also heating from friction. The rise in temperature tends to localize the deformation. However, in forming most sheet metals (aluminum, copper, and lower-strength steels), the localization caused by heating is a minor effect.

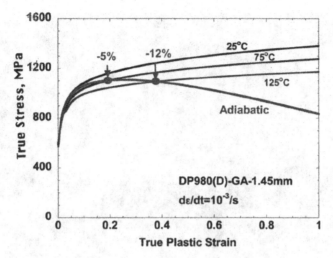

Figure 15.4. Adiabatic heating of a high-strength steel during forming can lead to early necking. From R. H. Wagoner, North American Deep Drawing Research Group presentation, May 4, 2010, Oakland University.

Adiabatic heating during bending of high-strength steels at high strain rates over sharp radii can lead to shear failures. Such a failure is illustrated in Figure 15.5.

TAYLOR-WELDED BLANKS

In recent years, parts are often stamped from blanks made by welding two or more sheets of different thickness or different base materials. The purpose of this is to save weight by using thinner-gauge material when possible and using thicker or stronger material only when necessary. Some difficulties are encountered during forming. Offset blank-holder surfaces are required to ensure adequate hold down. Welding hardens the weld zone, which may reduce the formability and cause cross-weld failures if the direction of major strain is parallel to the weld. The best blank orientation is with the weld perpendicular to the major strain axis. In this case, failure is likely to occur as a result of splitting of

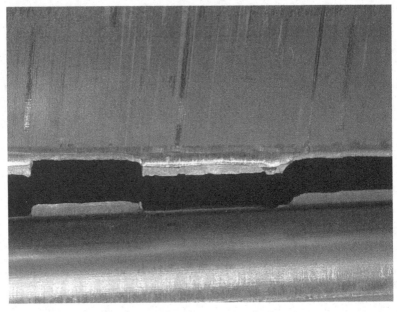

Figure 15.5. Shear failure of a high-strength steel during bending over a sharp radius. Courtesy of James Fekete, General Motors.

the thinner material parallel to the weld. This problem can be alleviated to some extent by decreasing the hold-down pressure on the thicker material and allowing more of the thicker metal to flow into the die. The frequency of failures parallel to the weld are reduced by decreasing the movement of the weld in the die.

SPECIAL SHEETS

Sandwich sheets with low-carbon steel on the outside around a polymer are used for sound dampening. For example, their use as the firewall between an automobile engine and the passenger compartment lowers engine noise.

Patterns can be impressed on the surface of sheets rolled with laser-textured rolls. It has been claimed that this permits better lubrication and better surface appearance after painting.

Figure 15.6. Spangles on the surface of a steel sheet that was hot-dip galvanized. From *The Making, Shaping* and Treating of Steel, 9th ed., U.S. Steel Corp. (1971).

SURFACE TREATMENT

Steel mills often sell prelubricated sheets or sheets coated with a polymer coating. Often steel is given a phosphate coating to help lubricants.

Steels are frequently galvanized (plated with zinc) for corrosion protection. Zinc is anodic to iron, and thus it galvanically protects the underlying metal. Steels may be galvanized by either hot dipping or electroplating. In the more common hot-dipping process, the thickness of the coating is controlled by wiping the sheet as it emerges from a molten zinc bath. Figure 15.6 shows the surface of a hot-dip galvanized sheet. In electroplating, the plating current and time control the thickness of electroplating. Usually the thickness of the zinc is the same on both sides of the sheet, but sheets can be produced with the thickness less on one side than the other. One-side-only plating is also possible.

The term *galvanneal* has been applied to hot-dip-galvanized sheets that are subsequently annealed to allow the formation of Fe-Zn intermetallic compound.

Other types of plating are sometimes done. Tin plating is an example, although "tin cans" today have little, if any, tin on them.

SPECIAL CONCERNS

Inclusion shape control: splitting of steels parallel to the rolling direction is sometimes a problem during fabrication or service. Usually the cracks form along elongated MnS particles. This causes the ductility measured in the transverse and through-thickness directions to be much lower than in the rolling direction. HSLA steels are much more sensitive to this than AKDQ steels. One common remedy is *inclusion shape control*. Additions of Ce and rare earths to the steel form sulfides that are much more resistant than MnS to elongation during hot rolling.

Hot shortness: copper and tin are called *tramp elements*. They enter steel through recycling of scrap. Unlike most other alloying elements, they are not oxidized during steel making. Their concentrations in steel are increasing from year to year because of repeated recycling. Trace amounts in steels can cause hot shortness. They are much less oxidizable than iron, so when the steel surface is oxidized during hot rolling, the concentrations of Cu and Sn can increase enough to melt.

Machinability of steel is improved by intentional additions of sulfur (resulfurized steels), which form MnS inclusions. Resulfurized steels must have increased manganese contents. Sometimes lead is added to hardenable steels for increased machinability.

REFERENCES

Flat Rolled Products, III, Metallugical Society Conference, v. 16, Interscience Publishers, Wiley (1962).

W. C. Leslie, *The Physical Metallurgy of Steels*, McGraw-Hill (1981).

D. T. Llewellyn and R. C. Hudd, *Steels: Metallurgy and Applications*, Butterworth Heinemann (1998).

The Making, Shaping and Treating of Steel, U.S. Steel Corp. (1971).

16

SHEET STEEL FORMABILITY

Low-carbon steel sheet is much more formable than aluminum sheet. Low-carbon sheet is formed by manufacturers into a variety of shapes, including automotive bodies, appliances, and cans. Formability and surface appearance are vital in most applications. Some of the important properties of sheet steels are the following:

1. surface finish
2. strain hardening
3. strain-rate sensitivity
4. anisotropy
5. freedom from yield point effect
6. yield strength

Surface finish: cold-rolled steels have a much a better surface finish than hot-rolled steels. The term *cold rolled* means the steel has been cold rolled and recrystallized (unless otherwise stated). Stretcher strains and orange peel (both discussed later in the chapter) are undesirable.

Strain hardening: the strain hardening is usually expressed by

$$\sigma = K\varepsilon^n, \tag{16.1}$$

where the strain-hardening exponent, n, describes the persistency of hardening and is equal to the uniform elongation in a tension test.

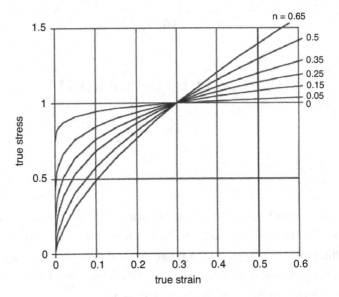

Figure 16.1. True stress-strain curves approximated by $\sigma = k\varepsilon^n$, with several values of n.

Figure 16.1 illustrates the relation between n and the shape of the stress-strain curve. A high n is indicative of a high stretchability.

Strain-rate sensitivity: the effect of the strain rate on the stress-strain curves is usually expressed by

$$\sigma = C\dot{\varepsilon}^m \tag{16.2}$$

for a constant level of strain. In steels with a high m, necks tend to localize less rapidly.

Anisotropy: this is usually as expressed by the plastic strain ratio

$$R = \varepsilon_w/\varepsilon_t, \tag{16.3}$$

where ε_w is the width strain and ε_t is the thickness strain in a tension test on a strip specimen (Figure 16.2). Generally high R leads to increased drawability (Figure 16.3) and less wrinkling.

For a discussion of freedom from strain aging, see Chapter 15.

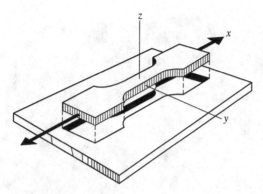

Figure 16.2. The R-value is the ratio of the width strain to the thickness strain. In an x-direction tension test, it is e_y/e_z. From W. F. Hosford and W. A. Backofen, in *Fundamentals of Deformation Processing; Proceedings of the 9th Sagamore AMRA Conference*, Syracuse University Press, Syracuse, NY (1964).

The steel-making practice affects many properties. High annealing (recrystallization) temperatures tend to raise n but produce a larger grain size. Large grain sizes promote a more pronounced orange-peel effect. This is the surface rumpling that occurs when a free surface is deformed (Figure 16.4). The roughness is on the scale of the grain size. With large grains the rumpling is so great that it won't be hidden by

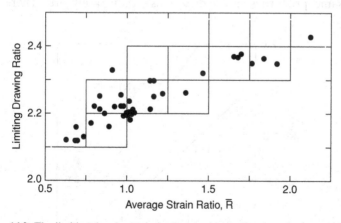

Figure 16.3. The limiting drawing ratio increases with the R-value. An increased limiting drawing ratio allows deeper cups to be drawn in a single operation. From R. W. Logan, D. J. Meuleman, and W. F. Hosford, in *Formability and Metallurgical Structure*, A. K. Sachdev and J. D. Embury, eds., TMS (1987).

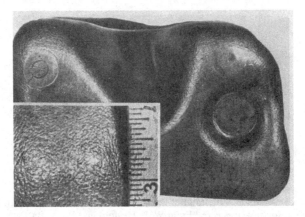

Figure 16.4. Orange peel. The surface rumpling is on the scale of the grain size. Courtesy of the American Iron and Steel Institute.

paint. Higher annealing temperatures also promote higher R-values as indicated in Figure 16.5.

ANISOTROPIC YIELDING

Although it is frequently assumed that materials are isotropic (have the same properties in all directions), they rarely are. There are

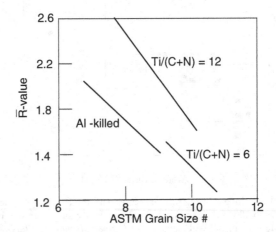

Figure 16.5. Increase of average R-value with increasing grain size. From D. J. Blickwede, *Metals Progress* (April 1969), p. 90.

two main causes of anisotropy. One cause is *preferred orientations* of grains, or *crystallographic texture*. The second is *mechanical fibering*, which is the elongation and alignment of microstructural features, such as inclusions and grain boundaries. Anisotropy of plastic behavior is almost entirely caused by the presence of preferred orientations.

The first complete quantitative treatment of plastic anisotropy was in 1948 by Hill,[*] who proposed an anisotropic yield criterion to accommodate such materials. It is a generalization of the von Mises criterion:

$$F(\sigma_y - \sigma_z)^2 + G(\sigma_z - \sigma_x)^2 + H(\sigma_x - \sigma_y)^2$$
$$+ 2L\tau_{yz}^2 + 2M\tau_{zx}^2 + 2N\tau_{xy}^2 = 1, \qquad (16.4)$$

where the axes x, y, and z are the symmetry axes of the material (for example, the rolling, transverse, and through-thickness directions of a rolled sheet).

If the loading is such that the directions of principal stress coincide with the symmetry axes and if there is planar isotropy (properties do not vary with direction in the x-y plane), Equation 16.1 can be simplified to

$$\left(\sigma_y - \sigma_z\right)^2 + (\sigma_z - \sigma_x)^2 + R(\sigma_x - \sigma_y)^2 = (R+1)X^2, \quad (16.5)$$

where X is the yield strength in uniaxial tension.

For plane stress ($\sigma_z = 0$) Equation 16.5 plots as an ellipse as shown in Figure 16.6. The higher the value of R, the more the ellipse extends into the first quadrant. Thus the strength under biaxial tension increases with R as suggested earlier.

The corresponding flow rules are obtained by applying the general equation, $d\varepsilon_{ij} = d\lambda(\partial f/\partial\sigma_{ij})$, where $f(\sigma_{ij})$ is given by Equation 16.5.

$d\varepsilon_x : d\varepsilon_y : d\varepsilon_z$

$$= [(R+1)\sigma_x - R\sigma_y - \sigma_z] : [(R+1)\sigma_y - R\sigma_x - \sigma_z] : [2\sigma_z - \sigma_y - \sigma_x].$$
$$(16.6)$$

[*] R. Hill, "Plastic Anisotropy," *Proc. Royal Soc.*, v. 193A (1948), p. 28.

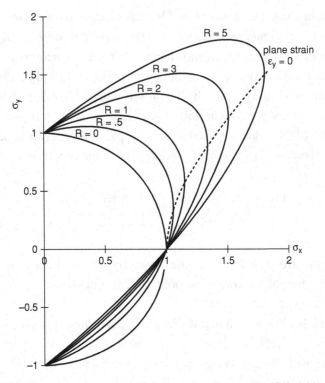

Figure 16.6. Plane stress ($\sigma_z = 0$) yields loci predicted by the Hill 1948 yield criterion for planar isotropy (Equation 16.5) with several values of R. The dashed line is the locus of stress states that produce plane strain ($\varepsilon_y = 0$). Note that the strength under biaxial tension increases with R. A high R indicates a resistance to thinning in a tension test, which is consistent with high strength under biaxial tension where thinning must occur. From W. F. Hosford, *The Mechanics of Crystals and Textured Polycrystals*. Used by permission of Oxford University Press, Inc.

This means that $\rho = \varepsilon_y/\varepsilon_x$ for $\sigma_z = 0$ and $\alpha = \sigma_y/\sigma_x$ for $\sigma_z = 0$ are related by

$$\rho = [(R+1)\alpha - R]/[(R+1) - R\alpha] \qquad (16.7)$$

and

$$\alpha = [(R+1)\rho + R]/[(R+1) + R\rho]. \qquad (16.8)$$

The effective stress

$$\bar{\sigma} = \{[(\sigma_y - \sigma_z)^2 + (\sigma_z - \sigma_x)^2 + R(\sigma_x - \sigma_y)^2]/(R+1)\}^{1/2}. \quad (16.9)$$

For $\sigma_z = 0$,

$$\bar{\sigma}/\sigma_x = \{[\alpha^2 + 1 + R(1 - \alpha)^2]/(R+1)\}^{1/2}. \quad (16.10)$$

The effective strain function is given

$$\bar{\epsilon}/\epsilon_x = (\sigma 1/\bar{\sigma})(1 + \alpha\rho). \quad (16.11)$$

The Hill criterion often overestimates the effect of R-value on the flow stress. A modification of Equation 16.2, referred to as the high-exponent criterion, was suggested to overcome this difficulty.[*]

$$(\sigma_y - \sigma_z)^a + (\sigma_z - \sigma_x)^a + R(\sigma_x - \sigma_y)^a = (R+1)X^a, \quad (16.12)$$

where a is an even exponent much higher than 2. Calculations based on crystallographic slip have suggested that a = 6 is appropriate for body-centered cubic metals and a = 8 for face-centered cubic metals. Figure 16.7 compares the yield loci predicted by this criterion and the Hill criterion for several levels of R.

With this criterion, the flow rules are as follows:

$$d\varepsilon_x : d\varepsilon_y : d\varepsilon_z$$

$$= [R(\sigma_x - \sigma_y)^{a-1} + (\sigma_x - \sigma_z)^{a-1}] :$$
$$[(\sigma_y - \sigma_z)^{a-1} + R(\sigma_y - \sigma_x)^{a-1}] :$$
$$[(\sigma_z - \sigma_x)^{a-1} + (\sigma_z - \sigma_y)^{a-1}]. \quad (16.13)$$

The effective stress function corresponding to Equation 16.9 is

$$\bar{\sigma} = [(\sigma_y - \sigma_z)^a + (\sigma_z - \sigma_x)^a + R(\sigma_x - \sigma_y)^a]/(R+1)\}^{1/a}. \quad (16.14)$$

[*] W. F. Hosford, Proceedings of the *7th North American Metalworking Conference*, SME, Dearborn MI (1979), pp. 191–197, and R. W. Logan and W. F. Hosford, *Int. J. Mech. Sci.*, v. 22 (1980), pp. 419–430.

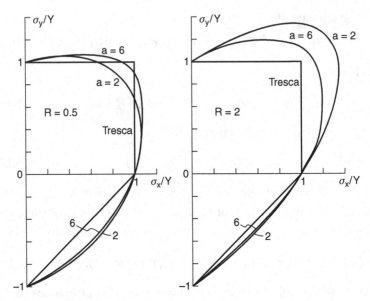

Figure 16.7. Yield loci predicted by the high-exponent criterion for planar isotropy. The loci for the 1948 Hill criterion correspond to a = 2. Note that with a high exponent, there is much less effect of R on strength under biaxial tension. From W. F. Hosford, ibid.

The effective strain function is

$$\bar{\varepsilon} = (\sigma_x/\bar{\sigma})(1 + \alpha\rho), \tag{16.15}$$

where $\alpha = (\sigma_y - \sigma_z)/(\sigma_x - \sigma_y)$ and $\rho = \varepsilon_y/\varepsilon_x$.

Although the nonlinear flow rules for the high-exponent criterion (Equation 16.10) cannot be explicitly solved for the stresses, iterative solutions with calculators or personal computers are simple. The nonquadratic yield criterion and accompanying flow rules (Equations 16.9 and 16.10) have been shown to give better fit to experimental data than the quadratic form (Equations 16.2 and 16.6).

EFFECT OF STRAIN HARDENING
ON THE YIELD LOCUS

According to the *isotropic hardening* model, the effect of strain hardening is simply to expand the yield locus without changing its shape.

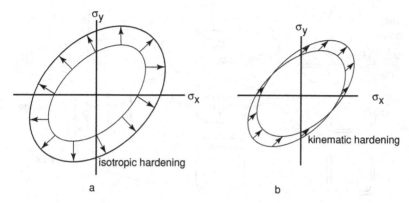

Figure 16.8. The effect of strain hardening on the yield locus. The isotropic model (a) predicts an expansion of the locus. The kinematic hardening model (b) predicts a translation of the locus in the direction of the loading path.

The stresses for yielding are increased by the same factor along all loading paths. This is the basic assumption that $\bar{\sigma} = f(\bar{\varepsilon})$. The isotropic hardening model can be applied to anisotropic materials. It does not imply that the material is isotropic.

An alternative model is *kinematic hardening*. According to this model, plastic deformation simply shifts the yield locus in the direction of the loading path without changing its shape or size. If the shift is large enough, unloading may actually cause plastic deformation. The kinematic model is probably better for describing small strains after a change in load path. However, the isotropic model is better for describing behavior during large strains after a change of strain path. Figure 16.8 illustrates both models.

DEEP DRAWING

Major concerns in sheet forming are wrinkling and tensile failure by necking. Sheet-forming operations may be divided into *drawing*, in which one of the principal strains in the plane of the sheet is compressive, and *stretching*, in which both of the principal strains in the plane of the sheet are tensile. Compressive stresses normal to the sheet are usually negligible in both cases.

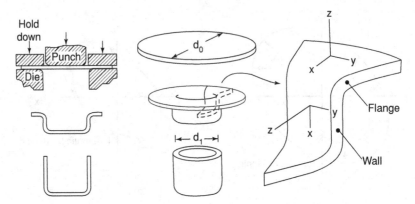

Figure 16.9. Schematic illustration of cup drawing showing the coordinate axes. As the punch descends, the outer circumference must undergo compression so that it will be small enough to flow over the die lip. From W. F. Hosford and R. M. Caddell, *Metal Forming: Mechanics and Metallurgy*, 4th ed., Cambridge University Press (2010).

A typical drawing process is the making of cylindrical, flat-bottom cups. It starts with a circular disc blanked from a sheet. The blank is placed over a die with a circular hole, and a punch forces the blank to flow into the die cavity, as sketched in Figure 16.9. A hold-down force is necessary to keep the flange from wrinkling. As the punch descends, the blank is deformed into a hat shape and finally into a cup. Deforming the flange consumes most of the energy. The energy expended in friction and some in bending and unbending as material flows over the die lip is much less. The stresses in the flange are compressive in the hoop direction and tensile in the radial direction. The tension is a maximum at the inner lip, and the compression a maximum at the outer periphery.

As with wire drawing, there is a limit to the amount of reduction that can be achieved. If the ratio of the initial blank diameter, d_0, to the punch diameter, d_1, is too large, the tensile stress required to draw the material into the die will exceed the tensile strength of the wall, and the wall will fail by necking. It can be shown that for an isotropic

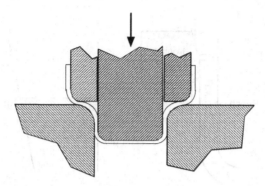

Figure 16.10. Direct redrawing. A sleeve around the punch acts as a hold down while the punch descends.

material, the highest ratio of d_o/d_1 that can be drawn (*limiting drawing ratio*, or LDR) is

$$(d_o/d_1)_{max} = \exp(\eta), \qquad (16.15)$$

where η is the deformation efficiency. For an efficiency of 70%, $(d_o/d_1)_{max} = 2.01$. There is little thickening or thinning of the sheet during drawing, so this corresponds to a cup with a height-to-diameter ratio of 3/4. Materials with R-values greater than unity have somewhat higher limiting drawing ratios. This is because with a high R-value, the increased thinning resistance permits higher wall stresses before necking as well as easier flow in the plane of the sheet, which decreases the forces required. Forming cylindrical cups with a greater height-to-diameter ratio requires *redrawing*, as shown in Figure 16.10. In can making, an additional operation called *ironing* thins and elongates the walls, as illustrated in Figure 16.11.

STAMPING

Operations variously called *stamping*, *pressing*, or even *drawing* involve clamping the edges of the sheet and forcing it into a die cavity by a punch, as shown in Figure 16.12. The metal is not squeezed

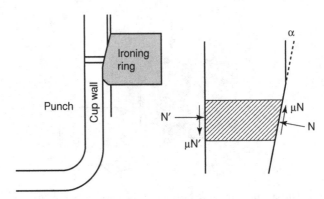

Figure 16.11. Section of a cup wall and ironing ring during ironing. The die ring causes wall thinning. Friction on opposite sides of the wall acts in opposing directions. From W. F. Hosford and R. M. Caddell, ibid.

between tools; rather it is made to conform to the shape of the tools by stretching. Failures occur by either *wrinkling* or *localized necking*. Wrinkling will occur if the restraint at the edges is not great enough to prevent excessive material being drawn into the die cavity. Blank-holder pressure and draw beads are often used to control the flow of material into the die. If there is too much restraint, more stretching may be required to form the part than the material can withstand.

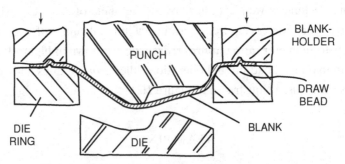

Figure 16.12. Sketch of a sheet-stamping operation by J. L. Duncan. The sheet is stretched to conform to the tools rather than being squeezed between them. In this case, the lower die contacts the sheet and causes a reverse bending only after it has been stretched by the upper die. For many parts, there is not a bottom die. From W. F. Hosford and R. M. Caddell, ibid.

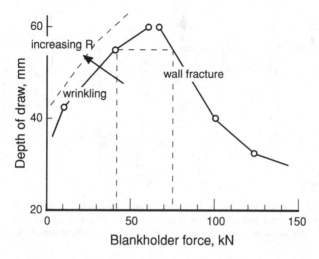

Figure 16.13. The effect of blank-holder force on the possible depth of draw. If the blank-holder force is too low, wrinkling will result from too much material being drawn into the die cavity. Too high a blank-holder force will require too much stretching of the sheet and result in a necking failure. For deep draws, there may be only a narrow window of permissible blank-holder forces. High *R*-values widen this window. From W. F. Hosford and R. M. Caddell, ibid.

The result is that there is a window of permissible restraint for any part, as illustrated in Figure 16.13. With too little blank-holder force, the depth of draw is limited by wrinkling. On the other hand, if the blank-holder force is too high, too little material is drawn into the die cavity to form the part, and the part fails by *localized necking*. The size of this window depends on properties of the sheet. Materials with higher strain-hardening exponents, n, can stretch more before necking failure, so the right-hand limit is raised and shifted to the left. There will be more lateral contraction in materials having a high *R*-value, thus decreasing the wrinkling tendency. Therefore, high *R*-values increase the wrinkling limit and shift it to the left.

FORMING LIMITS

In a tension test of a ductile material, the maximum load and *diffuse necking* occur when $\sigma = d\sigma/d\varepsilon$. For a material that follows a

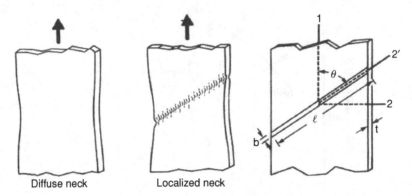

Figure 16.14. Development of a diffuse neck and a localized neck. The coordinate axes used in the analysis are shown in the last panel. The characteristic angle, θ, of the local neck must be such that $\varepsilon_2' = 0$. From W. F. Hosford and R. M. Caddell, ibid.

power-law ($\sigma = K\varepsilon^n$) stress-strain curve, this is when $\varepsilon = n$. This diffuse necking occurs by local contraction in both the width and thickness directions and is generally not a limitation in practical sheet forming. If the specimen is wide (as a sheet is), such localization must be gradual. Eventually a point is reached at which lateral contraction in the plane of the sheet ceases. At this point, a *localized neck* forms in which there is only thinning. In uniaxial tension, the conditions for localized necking are $\sigma = 2d\sigma/d\varepsilon$, or $\varepsilon = 2n$. Figure 16.14 illustrates general and localized necking in a tensile specimen. The characteristic angle at which the neck forms must be such that the incremental strain in that direction, $d\varepsilon_\theta$, becomes zero.

A plot of the combinations of strains that lead to necking failure is called a *forming limit diagram* (FLD). Figure 16.15 is such a plot for low-carbon steels. Combinations of strains below the forming limits are safe, whereas those above the limits will cause local necking. Note that the lowest failure strains correspond to plane strain, $\varepsilon_2 = 0$.

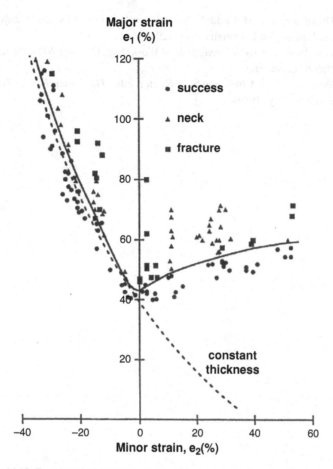

Figure 16.15. Forming limit diagram for low-carbon steel. The strain combinations below the curve are acceptable, whereas those above it will cause local necking. The limiting strains here are expressed as engineering strains, although true strains could have been plotted. Data from S. S. Hecker, "Simple Technique for Determining Forming Limit Curves," *Sheet Metal Ind.*, v. 52 (1975).

REFERENCES

R. Hill, *Mathematical Theory of Plasticity*. Oxford University Press (1950).

W. F. Hosford, *Mechanical Behavior of Materials*, 2nd ed., Cambridge University Press (2010).

W. F. Hosford, *Physical Metallurgy*, 2nd ed., Taylor and Francis (2010).

W. F. Hosford and R. M. Caddell, *Metal Forming: Mechanics and Metallurgy*, 4th ed. Cambridge University Press (2010).

Report on Advanced High Strength Steel Workshop, October 2006. Available at http://mse.osu.edu/.

R. H. Wagoner, K. S. Chan, and S. P. Keeler, eds., *The Minerals, Metals, and Materials Society* (1989).

17

ALLOY STEELS

AMERICAN INSTITUTE OF STEEL AND IRON AND SOCIETY OF AUTOMOTIVE ENGINEERS (AISI-SAE) DESIGNATION SYSTEM

This system uses four (occasionally five) digits to identify a steel composition. The last two digits indicate the carbon content in one-hundredth of a percent usually to the nearest .05%. For example, a 1020 steel contains between 0.018 and 0.23% C. For carbon content over 1%, five digits are used.

The first two digits indicate the major alloying elements as shown in Table 17.1.

The AISI-SAE-designated steels may contain incidental amounts of Cu ($\leq$0.35%), Ni ($\leq$0.25), Cr ($\leq$0.20%), or Mo ($\leq$0.06%).

EFFECT OF ALLOYING ELEMENTS

Alloys are used primarily to increase the hardenability of the steels. They have no behavior during tempering. Strong carbide formers tend to increase the hardness of tempered martensite, as shown in Figure 17.1. The net effect of alloys on the Vickers hardness is the sum of the contributions of each element. Molybdenum increases the time to form pearlite much more than the time to form bainite, so it is

Table 17.1. *Alloy type*

Designation	Meaning
10xx	Plain carbon (less than 1% Mn)
11xx	Plain carbon with added S or P
Plain carbon with high Mn (over 1%)	
40xx	0.7–0.9% Mn, 0.15–0.215% Mo
41xx	0.7–0.9% Mn, 0.8–1.1% Cr, 0.15–0.215% Mo
43xx	0.45–0.8% Mn, 1.65–2% Ni, 0.4–0.9% Cr, 0.2–0.3% Mo
44xx	0.45–0.9% Mn, 0.35–0.6% Ni, 0.2–0.3% Mo
50xx	0.75–1% Mn, 0.3–0.6% Ni
51xx	0.70–1% Mn, 0.7–1% Cr
52xx	0.25–0.45% Mn, 1.3–1.6% Cr
86xx	0.7–1% Mn, 0.4–0.7% Ni, 0.3–0.6% Cr, 0.15–0.3% Mo

Alloys containing boron, vanadium, and lead are designated as follows:
B denotes addition of boron as in 51B60 and other alloys; BV denotes addition of both boron and vanadium as in TS 43BV12 and other alloys; L denotes addition of lead as in 10L18, to increase the machinability.

used in steels in which bainite is a desired product. Alloying elements increases the times required for tempering. Mn, Ni, Cr, and Si tend to promote temper embrittlement; Mo, Ti, and Zr delay it.

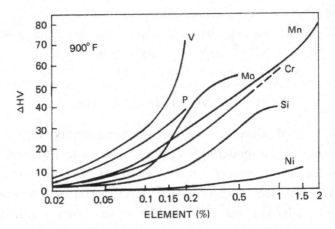

Figure 17.1. Effect of alloying elements on the hardness of martensite tempered for 1 h at 482°C. From R. A. Grange, C. R. Hibal, and L. P. Porter, "Hardness of Tempered Martensite in Carbon and Low Alloy Steels," *Met. Trans.*, v. 7A, pp. 1775–1785.

APPLICATIONS

Low-carbon alloys (0.10–0.25%C) are used primarily for carburized parts. These include 4023, 4118, and 5015. Intricate parts containing more than 0.40% C must be oil quenched to avoid cracking. The alloy 52100 is used exclusively for ball bearings. Springs are usually made from alloys 5155 and 5160.

REFERENCES

W. C. Leslie, *The Physical Metallurgy of Steels*, Hemisphere (1981).
The Making, Shaping and Treating of Steels, U.S. Steel Corp., 9th ed. (1971).
J. D. Verhoeven, *Steel Metallurgy for the Nonmetallurgist*, American Society for Metals (2007).

18

OTHER STEELS

HADFIELD AUSTENITIC MANGANESE STEEL

Manganese is a powerful austenite stabilizer as indicated by the Fe-Mn phase diagram (see Figure 8.2). Hadfield manganese steels containing 10 to 14% Mn and 1 to 1.4% C are austenitic at all temperatures. They are extremely wear resistant and are used in ore grinding and for teeth on earth-moving equipment. These steels work harden rapidly and as a consequence are difficult to machine. Parts are almost always cast to final shape.

MARAGING STEELS

Maraging steels develop high hardness by precipitation hardening of a very-low-carbon martensite. They contain about 18% Ni, 8 to 12% Co, and 4% Mo with very low carbon (<0.03%), as well as titanium (0.20 to 1.80%), and aluminum (0.10 to 0.15%). Their primary use is in tools and dies. Because the carbon content is so low, the martensite, which is formed by austenitizing at 850°C and cooling, is soft enough to be machined. Finished tooling can then be hardened by aging at 480°C for 3 hours. Whereas tools made from conventional tool steels have to be austenitized, quenched, and tempered after machining, maraging steels need only be heated to a moderate temperature to age and can be furnace cooled. This avoids oxidation, distortion, and cracking that often occurs during conventional heat treatment. The main disadvantage is the high cost that results from the high Ni, Mo, and Co contents.

TOOL STEELS

Most steel tools are not made from tool steels, but rather from carbon or low-alloy steels. Tool steels are used for shaping and cutting metal or rocks, wood, or concrete. The various classes are as follows:

W – *water hardening:* they are used for files, wood-working tools, drill bits, axes, and taps. They contain 0.6 to 1.4% C. Some have small amounts of Cr and V.

S – *shock resisting:* their uses include chisels, punches, and riveting tools. Typical compositions are 0.5% C, and Cr, Mo, Si 1 to 3%.

P – *mold steels:* the carbon contents are lower so that they can be machined and then carburized. Typical compositions are 0.10 to 0.30 C, 1.5 to 5 Cr, and 0.0 to 4 Ni. The lower carbon core gives them thermal shock resistance.

D – *cold working:* applications include cold-forming and thread-rolling dies. They have high carbon contents (1.5 to 2%) with typically 12% Cr and 1% Mo. They are air or oil hardening and have good dimensional stability.

H – *hot working:* these are used for forging dies, extrusion dies, and die-casting molds. Compositions range from 0.35 to 0.40% C, 3 to 5% Cr, 0.4 to 2% V, and either 1.5 to 2.5% Mo or 1.5 to 4% W. The latter elements ensure hot hardness.

M – *molybdenum-type high-speed steels:* these contain 0.8 to 1.1% C, 4 to 9% Mo, 1 to 2% W, 4% Cr, 1 to 2% V, and often Co.

T – *Tungsten-type high-speed steels:* these contain 0.7 to 1.8% C, 18% W, 4% Cr, 1 to 2%V, and often 5% or more Co.

The most important properties of tool steels are wear resistance, toughness, and hot hardness. Table 18.1 compares these properties in various grades of tool steel.

Table 18.1 *Relative values of important properties of tool steels*

Type	AISI	Wear resistant	Toughness (shallow hardened)	Hot hardness
Carbon	W1	4	7	1
Low alloy	L6	3	6	2
Shock resistant	S2	2	8	1
Die steels	O2	4	3	3
Cold working	A2	6	5	5
Die steels	H13	3	9	6
Hot working	H21	4	6	8
High speed	M2	7	3	8
	T1	7	3	8
	T15	9	1	9

AISI = American Institute of Steel and Iron.
From J. D. Verhoeven, *Steel Metallurgy for the Nonmetallugist*, ASM (2007).

Both molybdenum and tungsten high-speed tool steels rely on secondary hardening to retain a high hardness during use at the high temperatures that cutting tools see during high-speed machining. A number of carbides are precipitated during secondary hardening (Figure 18.1), and most of these are harder than the martensite from which they precipitate, as indicated in Figure 18.2.

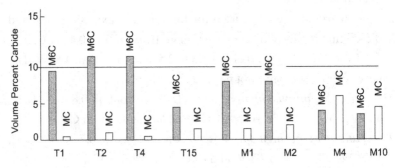

Figure 18.1. Carbides formed in T- and M-type tool steels after being annealed at the temperatures indicated. Data from F. X. Kayser and M. Cohen, *Metals Progress*, v. 61.

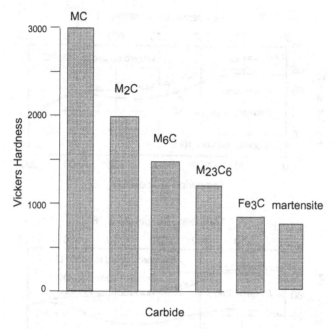

Figure 18.2. The hardnesses of various carbides. Data from E. Haberling, A. Rose, and H. H. Weigand, *Stahl und Eisen*, v. 93 (1973).

HEAT TREATMENT OF TOOL STEELS

Tool steels are austenized in the two-phase austenite-carbide temperature range. For steels containing high alloy contents, it may be necessary to preheat before austenitizing to eliminate temperature gradient that may cause cracking. Higher austenitizing temperatures lower the M_s temperature and increase the amount of retained austenite at room temperature. To eliminate retained austenite, it is common to subject quenched parts to subzero temperatures to increase the amount of martensite. The effect on hardness is illustrated in Figure 18.3.

The austenitizing temperature also affects the austenite grain size (Figure 18.4).

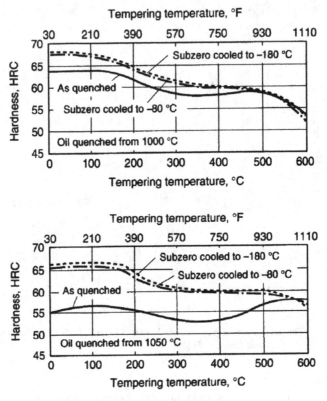

Figure 18.3. The effect of subzero cooling on the hardness of A2 tool steel after quenching. Note that the effect is greater after quenching from 1050°C than from 1000°C because of more retained austenite after quenching. From K.-E. Thelning, *Steel and Its Heat Treatment*, 2nd ed., Butterworth (1984).

For high-speed tool steels, multiple tempering is commonly employed, as indicated in Figure 18.5.

Tempering causes a secondary hardening in tool steels, as discussed in Chapter 14 and show in Figure 14.8.

NOTE OF INTEREST

Sir Robert Hadfield, in 1882, first developed Hadfield manganese steel. He was searching for a composition that would have a

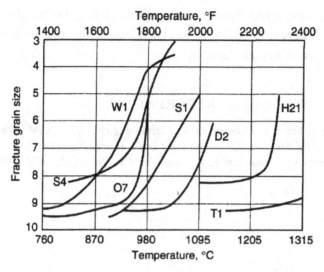

Figure 18.4. The dependence of austenite grain size as revealed by fracture on austeni-tizing temperature. From G. Roberts, G. Krauss, and R. Kennedy, *Tool Steels*, 5th ed., ASM (1998).

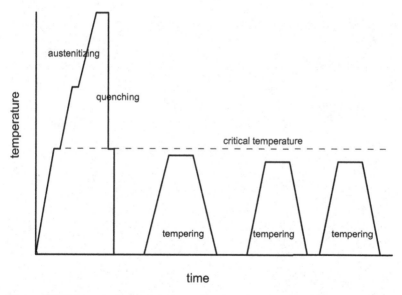

Figure 18.5. Schematic diagram of a heat treatment for a high-speed steel showing two preheat stages and triple tempering.

combination of high hardness and high toughness. The results of his first tests were so good that he had trouble believing them until he had made numerous other tests.

REFERENCES

G. Krauss, *Steels: Heat Treatment and Processing Principles*, ASM (1990).
W. C. Leslie, *The Physical Metallugy of Steels*, Hemisphere (1981).
G. Roberts, G. Krauss, and R. Kennedy, *Tool Steels*, 5th ed., ASM (1998).

19

STAINLESS STEELS

GENERAL CORROSION RESISTANCE

Stainless steels are characterized by a very good aqueous corrosion resistance and by a very good resistance to oxidation at high temperatures. All stainless steels contain at least 11% Cr. Many contain nickel as well. For the aqueous corrosion resistance, the steels must contain a minimum of 11.5% chromium, which makes them passive in oxidizing solutions. Even more chromium is required for passivity in nonoxidizing solutions. Unless the chromium content is sufficient for passivity, the corrosion resistance of stainless steels is similar to steels without any chromium. Table 19.1 is a galvanic series of alloys. It shows that stainless steels may occupy two positions corresponding to the active and passive conditions.

There are five major types of stainless steels: ferritic, martensitic, austenitic, duplex, and precipitation hardenable.

FERRITIC STAINLESS STEELS

These contain 11.5 to 30% chromium with minor amounts of silicon and manganese. The carbon content is kept as low as possible. There are several grades with a 4xx designation. As Figure 19.1 shows, with Cr > 12.7%, pure Fe-Cr alloys are ferritic (body-centered cubic or bcc) at all temperatures. Because they are bcc, ferritic stainless steels

Table 19.1. *Galvanic series in seawater**

Most anodic
Magnesium
Magnesium alloys
Zinc
Aluminum
Aluminum alloys
Low-carbon steels
Austenitic stainless steel (active)
Lead
Tin
Muntz metal
Nickel
Brass
Copper 70–30 cupro-nickel
Austenitic stainless steel (passive)
Most cathodic
Titanium

* From *ASM Metals Handbook*, 8th ed., v. 10 (1975).

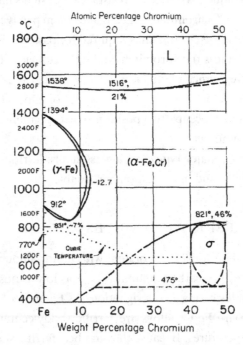

Figure 19.1. Iron-rich end of the Fe-Cr phase diagram. With chromium contents over 12.7%, stainless steels are ferritic at all temperatures. Sigma phase may develop at very high chromium contents, but the reaction is sluggish. From *ASM Metals Handbook*, 8th ed., v. 8, ASM (1973).

Table 19.2. *Ferritic stainless steels*

AISI No.	% Cr	% C	Other
405	11.5–14.5	0.08 max	0.1–0.3 Al
430	16–18	0.12 max	
446	23–27	0.20 max	

AISI = American Institute of Steel and Iron.

undergo a ductile-brittle transition at low temperatures. If C + N are kept below 0.015%, the transition temperature is below room temperature. Table 19.2 lists the common ferritic stainless steels.

The mechanical properties are similar to low-carbon steels. The strain-hardening exponent is about 0.20 and the R-value about 1. Major uses include architectural and automotive trim. A serious surface appearance may develop when sheets are stretched. Hills and valleys may form parallel to the prior rolling direction, as seen in Figure 19.2. This phenomenon, known as *roping* or *ridging*, is a result of bands of grains of two crystallographic textures, as illustrated in Figure 19.3. When a sheet is extended, grains with a {111}<112> orientation contract less laterally than grains with a {001}<110> orientation and therefore buckle.

If carbides precipitate during heat treatment, the stainless steels become *sensitized* and susceptible to intergranular corrosion, which is discussed in detail later. In ferritic stainless steels with a high chromium contents, sigma phase may form during heat treatment or in service at high temperatures. Figure 19.4 is an isothermal diagram for sigma phase formation.

In ferritic stainless steels with very high chromium contents, sigma phase may form during heat treatment or in service at high temperatures (Figure 19.1). This is regarded as undesirable because of the brittleness of this phase. Figure 19.4 is an isothermal transformation diagram for sigma phase formation.

Figure 19.2. Ridging on a ferritic stainless steel hubcap. From *The Making, Shaping and Treating of Steels*, 9th ed., U.S. Steel Corp. (1971).

MARTENSITIC STAINLESS STEELS

These contain 12 to 17% Cr, with 0.1 to 1.0% C. These also have (4xx) designations. The effect of C on the phases in a 12% Cr steel is shown in Figure 19.5. With 0.1% carbon, it is possible to form austenite. With 12% Cr, the hardenability is so great that rapid quenching is not

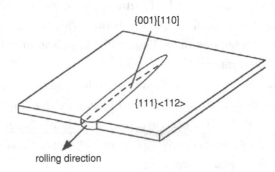

Figure 19.3. Combination of two textures that leads to ridging in ferritic stainless steels.

Table 19.3. *Martensitic stainless steels*

AISI No.	% Cr	% C	Other
410	11.5–13	0.15	
431	15–17	0.20	1.25–2.5 Ni
440A	16–18	0.65–0.75	0.75 Mo
440B	16–18	0.75–0.95	0.75 Mo
440C	16–18	0.95–1.2	0.75 Mo

AISI = American Institute of Steel and Iron.

necessary. The isothermal transformation diagram for a 12% Cr, 0.10% C steel (Figure 19.6) shows that about 3 minutes at 700°C are required to form the first pearlite. Martensitic stainless steels are used largely for razor blades, knives, and other cutlery. Table 19.3 lists several grades.

AUSTENITIC STAINLESS STEELS

These contain 17 to 25% Cr and 8 to 20% Ni. The amount of carbon is kept low. Austenitic stainless steels form the (2xx and 3xx)

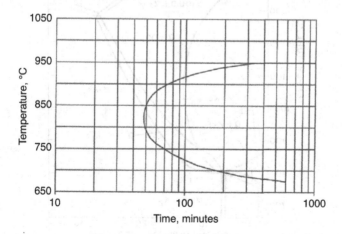

Figure 19.4. Isothermal transformation of sigma phase in a steel containing 25% Cr, 3% Mo, and 4% Ni. Data from E. L. Brown, M. E. Burnett, P. T. Purtscher, and G. Krauss, "Phase Formation in 25Cv-3Mo-4Ni Stainless Steel," *Met. Trans.*, v. 14A (1983), pp. 791–800.

Table 19.4. *Austenitic stainless steels*

AISI No.	% Cr	% Ni	C (max %)
302	18	9	0.15
304	19	9.3	0.08
304L	19	10	0.03
308	20	11	0.08
309	23	13.5	0.20
310	25	20.5	0.25

AISI = American Institute of Steel and Iron.

series (Table 19.4). Figure 19.7 is the isothermal section of the Fe-Cr-Ni ternary phase. The tendency of nickel to stabilize austenite is sufficient to overcome the ferrite-stabilizing tendency of chromium. It should be noted that 18% Cr 8% Ni falls just inside the γ region.

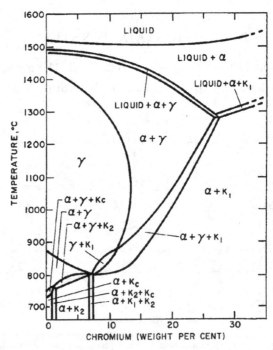

Figure 19.5. Section of the Fe-Cr-C ternary diagram at 10% C. Note that with 12% Cr, the steel can be heated into the austenite region. This makes it possible to form martensite. From *The Making, Shaping and Treating of Steel*, ibid.

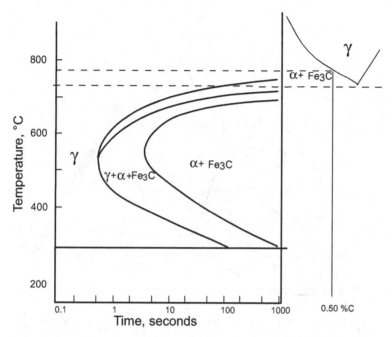

Figure 19.6. Isothermal transformation diagram for a 12% Cr, 0.10% C steel. Data from R. L. Rickett, W. F. White, C. S. Walton, and J. C. Butler, "Isothermal Transformation, Hardening and Tempering of 12% Chromium Steel," *Trans. ASM*, v. 44 (1952), pp. 138–175.

At lower temperatures, γ should transform to α, but the transformation is so sluggish that a steel containing 18% Cr and 8% Ni will remain austenitic at all temperatures. However, cold working may cause an austenitic stainless steel to transform by a martensitic reaction. Table 19.3 lists several grades of austenitic stainless steels.

Other elements affect the occurrence of austenite versus ferrite in stainless steels. Carbon and manganese tend to stabilize austenite, whereas Cr, Mo, Si, and Nb stabilize ferrite. The effects of these are characterized by nickel and chromium *equivalents*, as shown in Figure 19.8. Nitrogen is a strong austenite stabilizer and has been used to substitute for a portion of the nickel.

Austenitic stainless steels are resistant to aqueous corrosion over a larger range of conditions than ferritic stainless steels. Their mechanical properties include a very high strain-hardening exponent, *n* (in

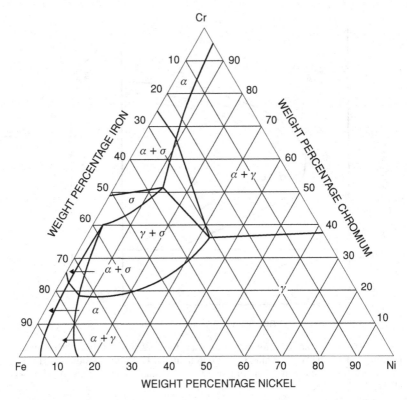

Figure 19.7. The 650°C isothermal section of the Fe-Cr-Ni ternary phase diagram. From *ASM Metals Handbook*, ibid.

the range of 0.5–0.65) and a low R-value, which is typical of face-centered cubic (fcc) metals. The high n can be attributed to the low stacking fault energy. Because they are fcc, they are not embrittled at low temperatures. One use for these stainless steels is in cryogenics. They are not nonferromagnetic, unless cold worked, enough to form martensite, so they can be identified by a magnet. They are useful in instruments that might be affected by magnetic fields.

OTHER STAINLESS STEELS

Precipitation-hardening stainless steels have an austenitic or a martensitic base with additions of Cu, Ti, Al, Mo, and Nb. Precipitates

Table 19.5. *Precipitation-hardening stainless steels*

AISI No.	Type	% Cr	% Ni	% C	Other
630	Martensitic	16	4.2	0.04	3.4 Cu, 0.25 Nb
631	Semiaustenitic	17	7.1	0.07	1.2 Al
600	Austenitic	14.8	25.3	0.05	low amounts of Mo, Al, Ti, V, B

AISI = American Institute of Steel and Iron.

include Ni_3Al, Ni_3Ti, $NiAl$, Ni_3Nb, and Ni_3Cu. Table 19.5 gives several compositions.

Duplex stainless steels a have mixed austenite-ferrite microstructure. Duplex stainless steels have a mixed microstructure of austenite and ferrite, the aim usually being to produce a 50/50 mix, although in commercial alloys the ratio may be 40/60. They are much stronger than austenitic stainless steels and have better resistance to pitting,

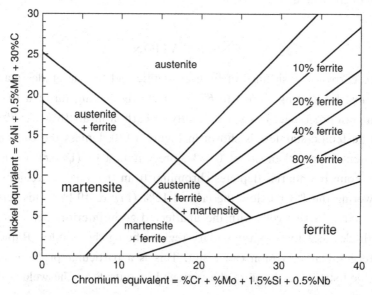

Figure 19.8. The nickel and chromium equivalents of several elements and their effects on the phases present. Data from *Metals Progress Data Book*, v. 112, ASM (1977).

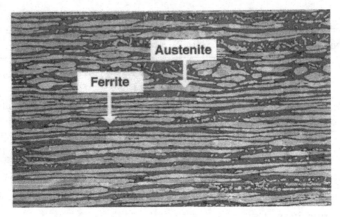

Figure 19.9. Two phase austenite-ferrite microstructure of a duplex stainless steel. From the International Molybdenum Association. Available at http://www.imoa.info/ moly_grade_steels/duplex–steel_html.

crevice corrosion, and stress corrosion cracking. The chromium content is high (19–28%) with lower nickel contents than austenitic stainless steels, and they contain up to 5% molybdenum. Figure 19.9 shows the microstructure.

SENSITIZATION

The corrosion resistance of both austenitic and ferritic grades may be lost by heating to 600 to 650°C or slowly cooling through that temperature range. Carbon solubility of carbon in an 18% Cr, 8% Ni austenitic stainless is shown in Figure 19.10. Unless the carbon content is less than about 0.03%, precipitation of $(Fe,Cr)_{23}C_6$ in the grain boundaries depletes chromium from the adjacent regions, lowering the %Cr below the critical 12% (Figure 19.11). This sets up active-passive cells with the unaffected grain interiors being the cathodes and the grain-boundary regions being the anodes. Rapid intergranular corrosion can result. This is a particular problem in welded structures because the regions just outside of the weld will

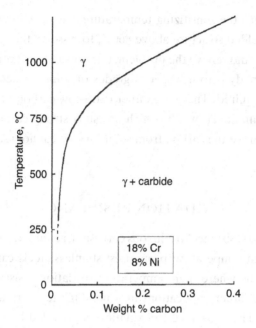

Figure 19.10. Carbon solubility in an 18–8 type stainless steel. From W. F. Hosford, *Physical Metallurgy*, 2nd ed., CRC Press (2010).

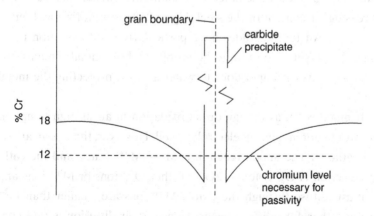

Figure 19.11. Precipitation of chromium carbide at the grain boundaries depletes the chromium in the adjacent regions to less than the critical 12% needed for passivity. From W. F. Hosford, ibid.

be heated into the sensitizing temperature range. Although heating the entire welded structure above 800°C to dissolve the carbides and quenching would relieve the problem, this is usually not practical. The common remedy is to use special grades of stainless steels in which welding is required. These are either extra-low-carbon grades or are grades containing Ti or Nb, which are such strong carbide formers that they remove the carbon from solution so it cannot react with the chromium.

OXIDATION RESISTANCE

In addition to resistance to aqueous corrosion, stainless steels are used for their high-temperature properties: stainless steels can be found in applications where high-temperature oxidation resistance is necessary and in other applications in which high-temperature strength is required. The resistance to oxidation is provided by an adherent Cr_2O_3 oxide film, which protects the underlying steel from oxidation. Oxidation requires diffusion of Cr^{+3} ions through the Cr_2O_3 film, which has few ionic defects. Oxidation resistance increases with increasing Cr content of the steel. Up to 26% is used for harsh environments. Austenitic grades have particularly good oxidation resistance. The layer is too thin to be visible, and the metal remains lustrous. The layer is impervious to water and air, protecting the metal beneath.

It might seem as though direct oxidation in air at high temperature would not involve an electrolytic cell. However, there is an anode and cathode. The anode reaction is $M \rightarrow M^{n+} + ne^-$ and the cathode reaction is $O_2 + 4e^- \rightarrow 2O^{2-}$. Either O^{2-} ions or M^{n+} ions and e^- must diffuse through the oxide. M^{n+} ions are smaller than O^{2-} ions and therefore diffuse faster. Hence, their diffusion is rate controlling. Figure 19.12 illustrates the reactions and transport in direct oxidation.

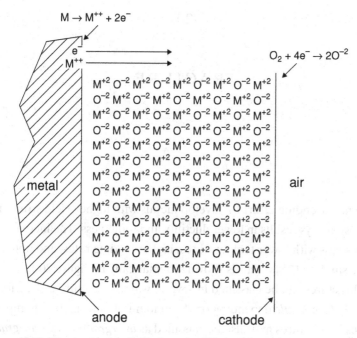

Figure 19.12. Direct oxidation. Oxide forms by diffusion of anions and electrons to the oxide-air surface.

REFERENCES

W. F. Hosford, *Physical Metallurgy*, 2nd ed., CRC Press (2010).
R. A. Lula, *Stainless Steel*, ASM (1986).
J. G. Parr and A. Hanson, *An Introduction to Stainless Steel*, ASM (1965).
J. D. Verhoeven, *Steel Metallurgy for the Nonmetallurgist*, ASM (2007).

20

FRACTURE

The never-ending effort to develop metals with higher yield strengths almost always results in lower ductility and toughness. The increase of flow stress with lower temperatures, higher loading rates, and notches have similar effects. This is shown schematically in Figure 20.1.

Fractures can be classified in several ways. A fracture is described as *ductile* or *brittle* depending on the amount of deformation that precedes it. Fractures may also be classified as *intergranular* or *transgranular*, depending on the fracture path. The terms *cleavage*, *shear*, and *void coalescence* describe the mechanism. These descriptions are not mutually exclusive; a brittle fracture may be either intergranular or transgranular.

DUCTILE FRACTURE

Fracture of a ductile steel in a tension test usually occurs by nucleation of voids at the center of the neck where the hydrostatic tension is greatest. These internal voids grow and eventually link up by necking of the metal between them. Figure 20.2 shows this schematically. Figure 20.3 is a section through a necked tensile specimen showing an internal crack formed by the linking of voids. Eventually, with the outward growth of the internal fracture, the outer rim can no longer support the load and will fail by sudden shear. The result is a cup-and-cone fracture (Figure 20.4). Voids form at inclusions because the

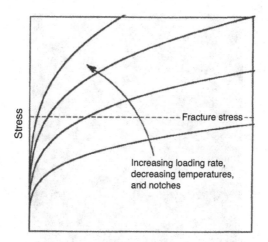

Figure 20.1. Lower temperatures, increased loading rates, and the presence of notches raise the flow stress so that the stress required for fracture is reached earlier. From W. F. Hosford and R. M. Caddell, *Metal Forming: Mechanics and Matallurgy*, 4th ed. Cambridge University Press (2011).

inclusion-matrix interface is weak or because the inclusion itself is weak. Figure 20.5 shows the fracture surface formed by void coalescence. As the number of inclusions increases, the distance between them decreases so that they link together at lower strains.

Mechanical working tends to produce directional microstructures. Grain boundaries and weak interfaces become aligned with the direction of working. Inclusions are elongated or broken up into strings of smaller inclusions. Often the loading in service is parallel to the

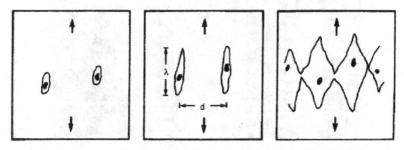

Figure 20.2. Schematic drawing showing the formation and growth of voids during tension and their linking up by necking. From W. F. Hosford, ibid.

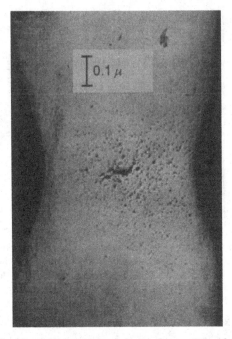

Figure 20.3. Section through a necked specimen of copper showing an initial crack formed by the linking of voids. From K. E. Puttick, *Phil. Mag.*, v. 4 (1959).

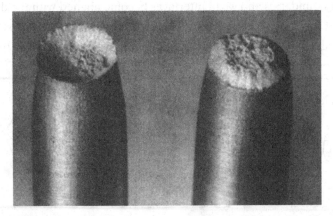

Figure 20.4. Typical cup-and-cone fracture in a tension test. From A. Guy, *Elements of Physical Metallurgy*, Addison-Wesley (1959).

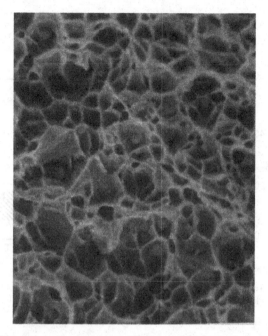

Figure 20.5. Dimpled ductile fracture surface in steel. Note the inclusions associated with about half of the dimples. The rest are on the mating surface. Courtesy of J. W. Jones. From W. F. Hosford and R. M. Caddell, ibid.

direction of alignment so that the alignment has little effect on ductility. For example, wires and rods are normally stressed parallel to their axes, and the stresses in rolled plates are normally in the plane of the plate. For this reason, the anisotropy caused by mechanical fibering is often ignored.

However, sometimes the largest stresses are normal to the aligned fibers. In these cases, fracture may occur by delamination. Welded T-joints may fail this way. In steels, elongated manganese sulfide inclusions are a major source of directionality. This effect can be eliminated by reducing the sulfur content and by adding small amounts of Ca, Ce, or Ti. These elements are stronger sulfide formers than manganese, and at hot-working temperatures, their sulfides are much harder than the steel, and so they do not become elongated. Figure 20.6 shows the effect of calcium on through-thickness ductility.

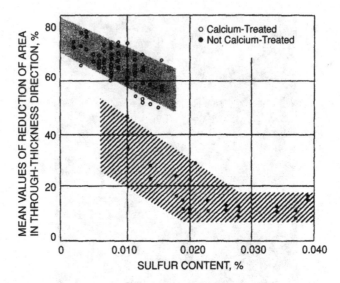

Figure 20.6. Effect of inclusion shape on through-thickness ductility of a high-strength low-alloy steel. From H. Pircher and W. Clapper, *Micro Alloying 75*, Union Carbide (1977).

The level of hydrostatic pressure plays a dominant role in controlling the strains at fracture. Hydrostatic compression tends to suppress void formation and growth. Photographs of specimens tested in tension under pressure (Figure 20.7) show that the reduction of area is greater under pressure. Figure 20.8 summarizes these data.

BRITTLE FRACTURE

One of the modes of brittle fracture is cleavage. Cleavage is the separation on atomic planes with little deformation. In iron, like all body-centered cubic metals, cleavage occurs on the {100} planes. It is believed that cleavage occurs when the stress normal to the cleavage plane reaches a critical value, σ_c. The critical applied stress, σ_a, is then

$$\sigma_a = \sigma_c/\cos^2 \phi, \tag{20.1}$$

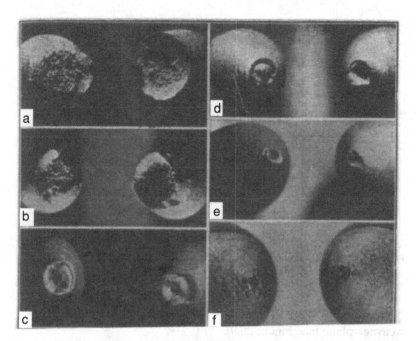

Figure 20.7. Effect of pressure on the area reduction in a tension test: (a) atmospheric pressure, (b) 234 kPa, (c) 1 MPa, (d) 1.3 MPa, (e) 185 MPa, and (f) 267 MPa. From P. W. Bridgman, in *Fracture*, ASM (1947).

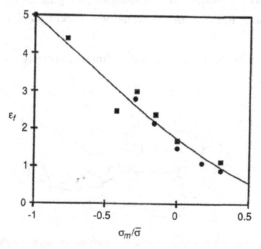

Figure 20.8. The effective strain at fracture of two steels increases as the mean stress, σ_m, becomes more negative. From W. F. Hosford and R. M. Caddell, ibid.

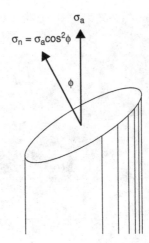

Figure 20.9. Cleavage occurs when the stress normal to the cleavage plane, $\sigma_n = \sigma_a \cos^2 \phi$, reaches a critical value, σ_c. From W. F. Hosford, ibid.

where ϕ is the angle between the tensile axis and the normal to the cleavage plane (see Figure 20.9).

In three dimensions, the cleavage plane in one grain will not be aligned with that in the neighboring grain, and so fracture cannot occur by cleavage alone. This is illustrated in Figure 20.10.

A cleavage fracture in an iron-3.9% nickel steel is shown in Figure 20.11.

Brittle fractures may also occur intergranularly as shown in Figure 20.12.

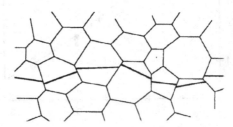

Figure 20.10. In a polycrystalline material, the cleavage planes in neighboring grains are not aligned. Another mechanism is required to link up cleavage fractures in neighboring grains. From W. F. Hosford, ibid.

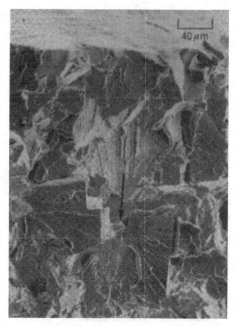

Figure 20.11. Cleavage fracture in a 3.9% Ni steel. Arrow indicates the direction of fracture. From *ASM Metals Handbook*, 8th ed., v. 9, ASM (1974).

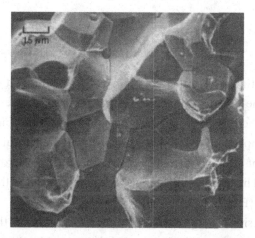

Figure 20.12. Intergranular fracture of pure iron under impact. From *ASM Metals Handbook*, ibid.

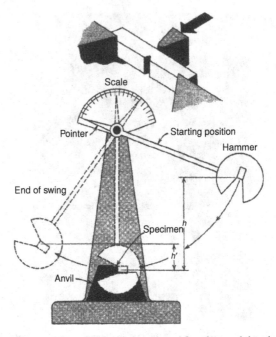

Figure 20.13. Charpy testing machine and test bar. After the pendulum breaks the bar, the height of the pendulum swing indicates the energy absorbed. From H. W. Hayden, W. G. Moffatt, and J. Wulff, *Structure and Properties of Materials, v. 3. Mechanical Behavior*, Wiley (1965).

TRANSITION TEMPERATURE

If a material absorbs much energy when it fractures, it is regarded as tough. The *Charpy* impact test is one method of assessing toughness. A notched bar is broken by a swinging pendulum (Figure 20.13). When the impact energy is plotted against the temperature of testing, it is apparent that the energy drops rapidly over a narrow temperature range. The middle of this range is commonly called the *transition temperature* (Figure 20.14).

The transition temperature of a steel depends on the criterion used, the sharpness of the notch, and the rate of loading and thus is not an inherent property. This is illustrated in Figure 20.15.

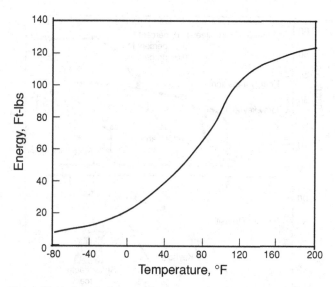

Figure 20.14. Ductile-brittle transition in a hot-rolled low-carbon steel.

Finer grain sizes and increased manganese and nickel contents lower the transition temperature, whereas increased carbon raises it (Figure 20.16).

Brittle fracture can occur without notches or rapid loading at very low temperatures. Figure 20.17 shows the fracture transition for very pure iron. Note that below the transition, fracture occurs on yielding.

LIQUID METAL EMBRITTLEMENT

Steels may be embrittled by exposure to liquid metals such as lead, solder, cadmium, several copper alloys, zinc, and lithium. The embrittlement is not time dependent and fracture may occur at very low stresses.

HYDROGEN EMBRITTLEMENT

High-strength steels can be embrittled by hydrogen. Two characteristics of hydrogen embrittlement are (1) Fractures are not immediate.

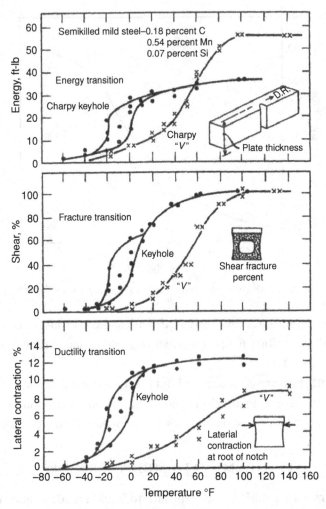

Figure 20.15. Charpy test results for two notches. The three criteria – energy absorption, fracture appearance, and deformation below the notch result in different transition temperatures. From Pellini, *Spec. Tech. Publ.,* v. 158 (1954).

They occur sometime after application of the load and hence are sometimes called *static fatigue*. (2) In a tension test, there is a lower strain to fracture. This loss of ductility increases with the amount of dissolved H and with the strength level of steel (Figure 20.18).

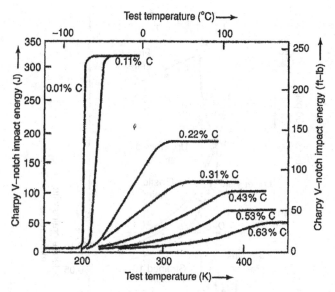

Figure 20.16. Effect of carbon content on Charpy V-notch energy. Increasing the carbon content raises the transition temperature and lowers the shelf energy. Data from J. A. Rinehbilt and W. J. Harris, *Trans. ASM*, v. 43 (1951).

Figure 20.19 shows a missile casing that failed by hydrogen embrittlement. Ductility is lower at low strain rates than high strain rates.

The sources of hydrogen in steel include pickling (H_2SO_4, HCl), electroplating (Cd, Cr, Zn), corrosion (esp. NaOH), welding (especially with coated electrodes), and exposure to H_2 at high temperatures. Hydrogen dissolves monatomically and diffuses rapidly to regions of high hydrostatic tension (notches). Hydrogen can be baked out of steel, as indicated in Figure 20.20.

FATIGUE

Steel, like all other metals, will fail at stresses well below the tensile strength after repeated cycles of loading and unloading. This phenomenon is called *fatigue*. Figure 20.21 is an "S-N" curve for a 4340

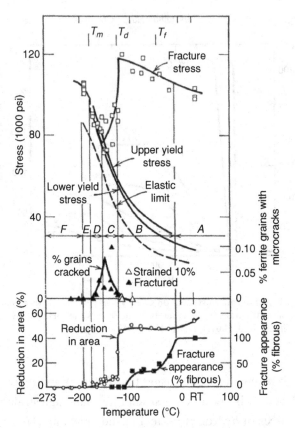

Figure 20.17. Fracture transition in very pure iron. From G. T. R. Hahn, B. L. Averbach, W. S. Owen, and M. Cohen, "Initiating of Cleavage Microcracks in Iron and Steel," in *Fracture*, Technology Press (1959).

steel. ("S-N" is short for cyclic stress vs. number of cycles to failure.) The S-N curve for steels usually has an endurance limit or fatigue strength, which is a stress level below which an infinite number of cycles can be applied without failure. The breaks in the curves usually occur near 10^6 cycles. This is in contrast to most other metals, which have no fixed endurance limit. Rather, the number of cycles to failure increases as the stress in lowered but never reaches a plateau.

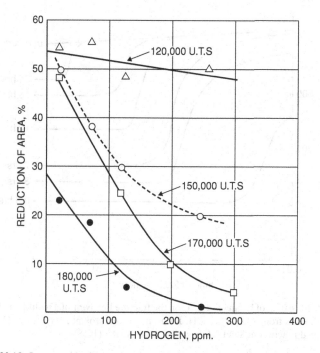

Figure 20.18. Increased hydrogen levels and increased steel strength result in a greater loss of ductility. From H. M. Burke et al., *Metals Prog.*, v. 67, no 5. (1955).

Figure 20.19. Remains of a missile casing after failure by hydrogen embrittlement. From Shank et al., *Metals Prog.*

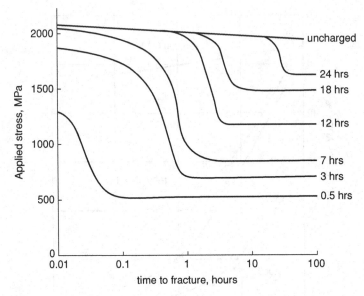

Figure 20.20. Effect of baking at 150°C on the fracture behavior of a hydrogen-charged 4340 steel. Data from J. O. Morlet, H. Johnson, and A. Troiano, "Hydrogen Cracking and Delayed Fracture in Steel," *J. Iron. Steel Inst.*, v. 189 (1958).

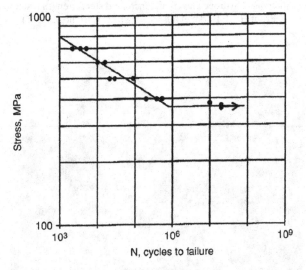

Figure 20.21. The S-N curve for a 4340 steel. W. F. Hosford, *Mechanical Behavoir of Material*, 2nd ed., Cambridge University Press (2010).

REFERENCES

B. L. Averbach, K. Feldbeck, G. T. Hahn, and D. A.Thomas, *Fracture*, Technology Press (1959).

P. W. Bridgman, *Fracture of Metals*, American Society for Testing and Materials (1947).

Fracture of Engineering Materials, ASM (1959).

W. F. Hosford, *Mechanical Behavior of Materials*, 2nd ed., Cambridge University Press (2010).

W. F. Hosford, *Physical Metallurgy*, 2nd ed., CRC Press (2010).

Metals Handbook, 8th ed., v. 9, ASM (1974).

21

CAST IRONS

PRODUCTION

Cast iron is produced by melting pig iron, often with substantial quantities of scrap iron and scrap steel, which lower the carbon and silicon contents to the desired levels, which may be anywhere from 2 to 4% and 1 to 3%, respectively. Other elements are often added in the form of ferro alloys to the melt before casting. In the past, most cast iron was melted in a cupola, which resembles a small blast furnace. Cold pig iron and steel scrap were charged from the top onto a bed of hot coke through which air was blown. Today, because of strict emission controls, most cast iron is produced in induction furnaces rather than cupolas.

GENERAL

Cast irons are a family of ferrous metals with a wide range of properties produced by being cast into shape as opposed by being formed. Figure 21.1 is the iron-carbon phase diagram. In the iron-carbon system, true equilibrium is between iron and carbon. Fe_3C is not an equilibrium phase. In steels, the carbon content is low enough that graphite rarely forms. In contrast, most cast irons have sufficient carbon that graphite does form. Note that the solubility of carbon in α and γ is lower when they are in equilibrium with graphite than in metastable equilibrium with cementite. The reason for this is

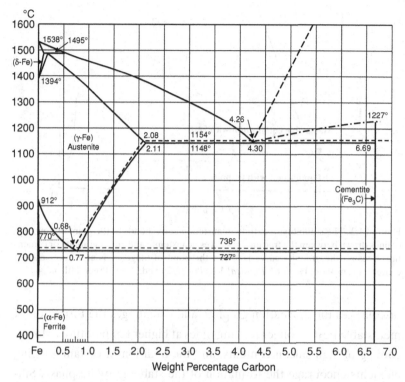

Figure 21.1. The iron-rich end of the iron-carbon phase diagram. Note that Fe$_3$C is a metastable phase. True equilibrium exists between γ and graphite and between α and graphite. The dashed lines are the true equilibrium with graphite, and the solid lines are the metastable equilibrium with cementite. From J. Chipman, *ASM Metals Handbook*, 8th ed., v. 8, ASM (1973).

illustrated in Figure 21.2. Other elements are used to control specific properties. Cast irons have lower melting temperature and material costs than steels and so are less expensive than cast steel. Cast irons have a wide range of mechanical properties, which make them suitable for use in engineering components. The widespread use of cast irons results from their low cost and versatile properties. An iron-carbon alloy containing more than 2% C cooled very slowly will solidify to graphite (pure carbon) and an iron-rich phase. At normal cooling rates, the metastable cementite Fe$_3$C is formed. Rapid cooling

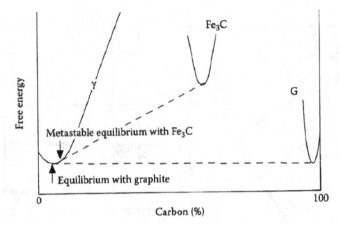

Figure 21.2. The solubility of carbon in austenite corresponds to the point of tangency of lines drawn between the free energy curves of austenite and either graphite or cementite. Because the graphite curve is lower, the point of tangency is at a lower carbon content. From W. F. Hosford, *Physical Metallurgy*, 2nd ed., CRC Press (2010).

discourages the nucleation graphite and encourages the formation of metastable Fe_3C. Longer holding times at higher temperatures, slower cooling of high-silicon contents, and the addition of certain alloying elements encourage the formation of the stable graphite phase. Silicon also raises the Fe-graphite eutectoid temperature while lowering that of Fe-cementite (Figure 21.3). Silicon shifts the eutectic to lower carbon contents. The percent carbon at the eutectic composition can be approximated by

$$\%C = 4.3 - 0.3\% \text{ Si.} \qquad (21.1)$$

The formation of graphite instead of Fe_3C is favored by:

1. high-carbon contents,
2. high-silicon contents,
3. long times at high temperatures or slow cooling, and
4. the absence of carbide formers (Cr, Mo, etc.).

There are four basic types of cast iron: white, gray, ductile, and malleable.

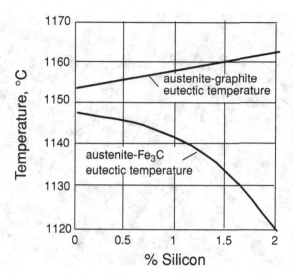

Figure 21.3. The effect of silicon on the eutectoid temperature. Data from I. C. H. Hughs in *Solidification of Metals*, ISTM (1967).

WHITE IRONS

White cast irons tend to have less carbon and silicon than the other grades – typically 2.5% C and 0.5% Si – so Fe_3C forms instead of graphite. The microstructure (Figure 21.4) contains massive carbides together with pearlite. White cast irons are brittle, with fracture occurring along paths of carbides. Their use is limited to special applications requiring high wear resistance such as tappets for automobile engines. Another use is as a precursor for the formation of malleable cast iron. The centers of large castings may not solidify white because the cooling rate is not fast enough. However, chromium additions allow white iron to be formed in large castings.

GRAY IRONS

These have sufficient carbon and silicon (typically 3.2% C and 2.5% Si) that graphite forms during freezing. The graphite occurs in the form of flakes, as shown in Figure 21.5. The three-dimensional

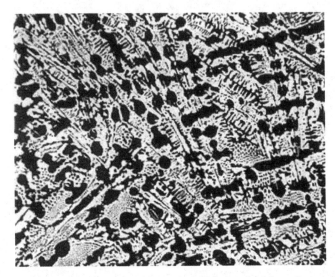

Figure 21.4. Microstructure of a white cast iron. Primary dendrites of austenite surrounded by an austenite-carbide eutectic formed during solidification. The austenite transformed to pearlite later. From *ASM Metals Handbook*, 8th ed., v. 7, ASM (1972).

Figure 21.5. Microstructure of a gray cast iron, unetched. The flakes of graphite are apparent. From *ASM Metals Handbook*, 8th ed., v. 7, ibid.

Figure 21.6. An electron micrograph of a gray iron that has been deeply etched to remove the ferrite and pearlite. The three-dimensional structure of the graphite flakes is apparent. From *ASM Metals Handbook*, 8th ed., v. 7, ibid.

nature of the graphite flakes is apparent in Figure 21.6. Gray cast irons are not very ductile. One percent elongation in a tension test is typical. Fracture follows a path through graphite flakes so that the fracture surface is almost entirely graphite. The fracture appearance is the origin of the term *gray iron.*

Gray cast iron is an inexpensive material. It is much easier to cast than steel because of its lower melting point and because there is almost no liquid-to-solid shrinkage. It is easy to machine because the graphite flakes allow chips to break into small pieces. Gray cast iron has a high capacity to dampen vibrations because of the graphite flakes. This is desirable in such applications as bases for lathes, milling machines, and other types of machinery that are sensitive to vibrations. Because of its low cost, it finds wide use for

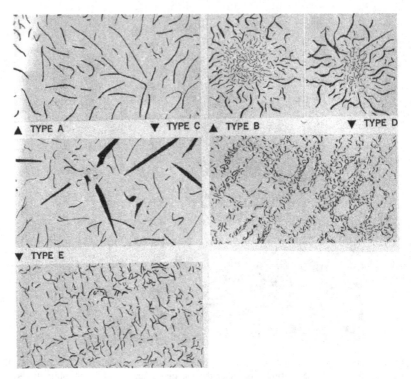

Figure 21.7. Types of graphite flakes. From *Gray and Ductile Iron Castings Handbook*, C. F. Waltron ed., Gray and Ductile Iron Founder's Society (1971).

manhole covers, fire hydrants, sewer gratings, drainage pipe, and lamp posts.

The tensile strength of gray iron decreases with the amount of flake graphite, and increased carbon and silicon contents lower the tensile strength. On the other hand, most alloying elements increase tensile strength. Slower cooling promotes larger graphite flakes. The shape of graphite flakes is classified as shown in Figure 21.7. The flake size is described by the length of the longest flakes, and flakes tend to be longer with higher carbon contents and slower cooling. Compressive strengths are much higher than tensile strengths (Figure 21.8).

Table 21.1 gives the classification system for gray cast iron.

Table 21.1. *Classification of gray cast iron*

Class	Carbon (%)	Silicon (%)	Tensile strength (psi)
20	3.40–3.60	2.30–2.50	20,000
30	3.10–3.30	2.10–2.30	30,000
40	2.95–3.15	1.70–2.00	40,000
50	2.70–3.00	1.70–2.00	50,000
60	2.50–2.85	1.70–2.00	60,000

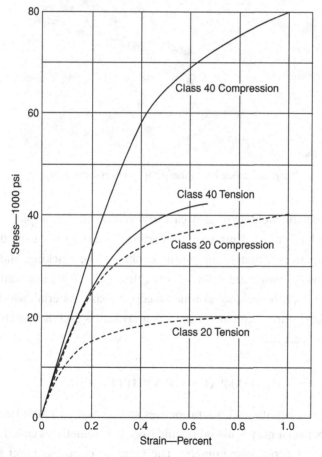

Figure 21.8. Difference between tensile and compressive strengths of gray iron. From C. F. Walton, ibid.

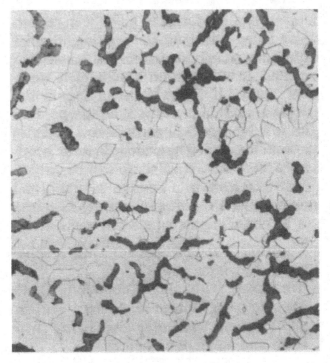

Figure 21.9. Compact graphite iron. From *ASM Metals Handbook*, 9th ed., v. 15, ASM (1988).

Successful welding of gray cast iron is difficult. The rapid cooling usually leads to the formation of martensite, further reducing ductility. The stresses that set up during cooling cause cracking, and thus very slow cooling after welding is required. The usual alternative to welding is braze welding in which a copper-base material, which has a melting temperature under the lower critical, is used to join the two pieces of cast iron.

COMPACT GRAPHITE IRON

Compact graphite iron has properties and a microstructure intermediate between gray iron and ductile iron. It is sometimes called *seminodular* or *vermicular graphite*. The graphite occurs as blunt flakes (Figure 21.9), a microstructure that results in properties intermediate

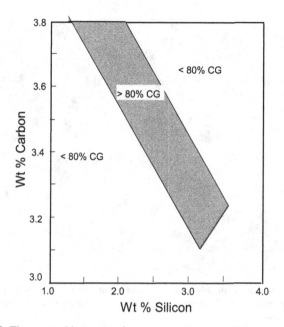

Figure 21.10. The composition range of compact graphite iron. Data from *ASM Metals Handbook*, 9th ed., v. 15, ASM (1988).

between gray and ductile irons. Production of compact graphite iron is similar to that of ductile iron, requiring close metallurgical control and rare earth element additions. An alloying element such as titanium is necessary to prevent the formation of spheroidal graphite. Compact graphite iron has much of the castability of gray iron but with tensile strengths and stiffness up to 75% higher. It is widely used for automobile engine blocks.

The carbon equivalent (CE) at which compact graphite forms is between 3.7 and 4.7, where CE is defined as

$$CE = \% \ C + (\% \ Si + \% \ P)/3.$$

Figure 21.10 shows this composition range. If the CE is above this range, the graphite may float; if it is lower than this range, chilling to form white iron may occur.

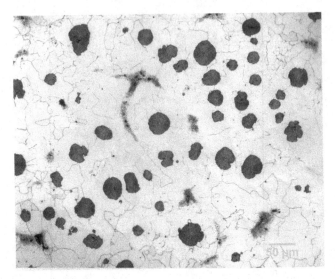

Figure 21.11. A ferritic ductile cast iron. Note that the graphite is in the form of spheroids, in contrast to the flake form in gray iron. Courtesy of Kathy Hayrynen, Applied Process, Inc.

DUCTILE CAST IRON

Ductile cast iron was patented in 1949. It has about the same compositions as gray iron except that the sulfur content is very low. Magnesium (or Ce and rare earths) is added to the melt just before pouring to further reduce the sulfur. The result is that the graphite forms as spheroids or nodules as shown in Figure 21.11. The final product has considerable toughness and ductility. It is sometimes called semisteel. It has replaced forged steel in applications such as automobile crankshafts.

Figure 21.12 shows several ranges of nodularilty. If the nodularity is low, the strength suffers (Figure 21.13).

MALLEABLE CAST IRON

Heating a white cast iron to just below the eutectic temperature (900°C) allows the iron carbide to decompose, $Fe_3C \rightarrow \alpha + graphite$.

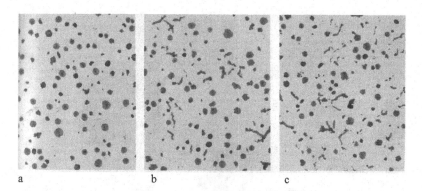

Figure 21.12. Nodularity of ductile cast irons: (a) 99%; (b) 80%; (c) 50%. From *ASM Metals Handbook*, 9th ed., v. 15, ASM (1988).

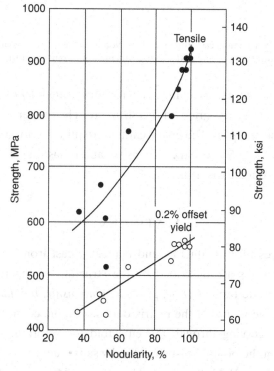

Figure 21.13. Tensile and yield strengths as a function of visually observed nodularity From A. G. Fuller, P. J. Emerson, and G. F. Sergeant, "A Report on the Effect Upon Mechanical Properties of Variation in Graphite Form in Irons Having Varying Amounts of Ferrite and Pearlite in the Matrix Structure and the Use of ND Tests in the Assessments of Mechanical Properties of Cast Irons," *Trans. AFS*, v. 88 (1980).

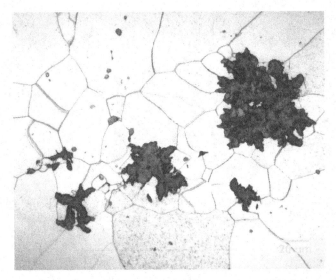

Figure 21.14. A malleable cast iron, unetched. Note that the graphite is in the form of rosettes, in contrast to the spheroidal form in ductile iron. Courtesy of Buck Co.

The graphite forms as clumps or rosettes called *temper carbon*. Figure 21.14 is a micrograph of this structure. The properties are similar to ductile cast iron. The heat treatment required to form malleable iron is long (2 to 5 days) and expensive, and it has thus been largely replaced by ductile cast iron.

MATRICES

The matrices of gray, ductile, and malleable cast iron is essentially steel. Cast irons are described by the structure of the matrix, for example, by the terms *ferritic, pearlitic, martensitic, bainitic*, or even *austenitic*. The nature of the matrix depends on the composition and the rate of cooling through the eutectoid temperature. It may be altered from the original cast structure by subsequent heat treatment. Figures 21.15 to 21.20 shows several cast iron microstructures.

Austempering of ductile cast iron produces a microstructure of acicular ferrite in a carbon-stabilized austenite matrix rather than the

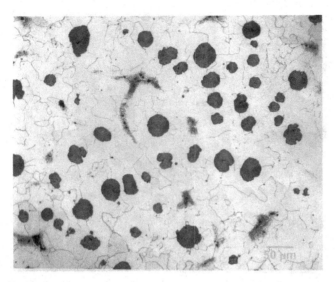

Figure 21.15. A ferritic ductile cast iron. Courtesy of Kathy Hayrynen and John Keough, Applied Process, Inc.

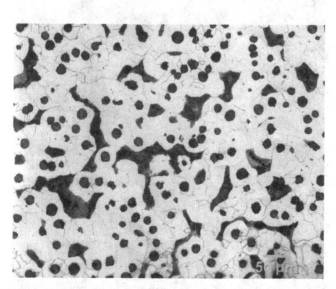

Figure 21.16. A ferritic ductile cast iron with some pearlite (dark areas). Courtesy of Kathy Hayrynen and John Keough, Applied Process, Inc.

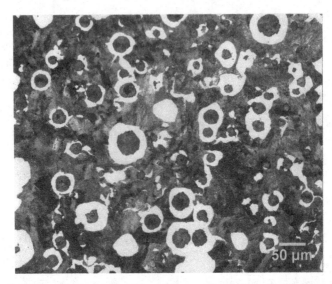

Figure 21.17. A pearlitic ductile cast iron with some ferrite. Note that the ferrite is adjacent to the graphite. Courtesy of Kathy Hayrynen and John Keough, Applied Process, Inc.

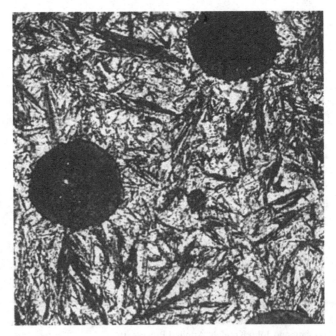

Figure 21.18. A martensitic ductile cast iron. From *ASM Metals Handbook*, 8th ed., v. 7, ibid.

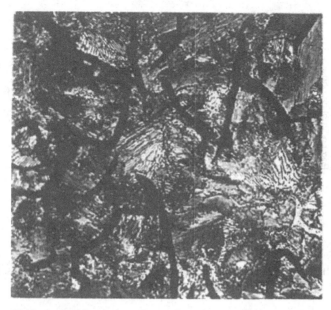

Figure 21.19. A pearlitic gray iron. From *ASM Metals Handbook*, 8th ed., v. 7, ibid.

Figure 21.20. Martensitic malleable cast iron. The matrix is tempered martensite. Courtesy of Buck Co.

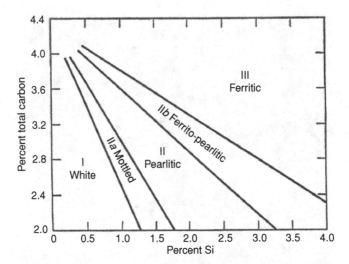

Figure 21.21. The effect of composition on the microstructure of sand-cast bars. From J. Wulff, H. F. Taylor, and A. J. Shaler, *Metallurgy for Engineers*, Wiley (1952).

bainite found in austempered steels. The properties can be varied with heat treatment. Very high tensile strengths can be achieved along with very good toughness.

Figure 21.21 summarizes the effects of composition on the microstructure for cast irons in the sand-cast condition. Figure 21.22 summarizes how composition and heat treatment control the phases in cast irons.

AUSTEMPERING OF CAST IRONS

Austempering of cast irons produces a microstructure different from austempered steel. The compositions are 3.6 to 3.8% C, 2.4 to 2.8% Si, and less than 0.3% Mn. The properties can be varied by changing the austempering temperature from a yield strength of 500 MPa for 400°C to 1400 MPa at 260°C.

The first commercial austempering of ductile cast iron occurred in 1972, and by 1998, it had become an important product with a

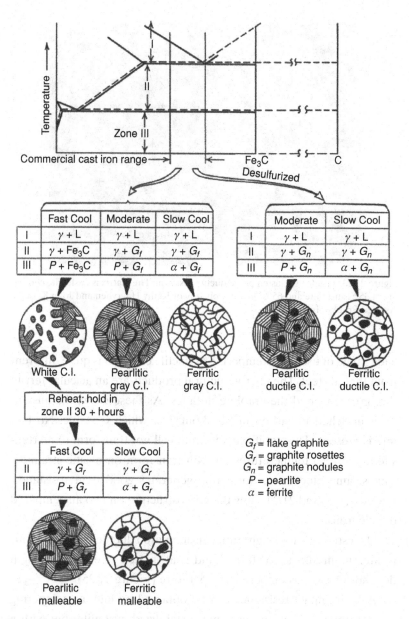

Figure 21.22. Effect of composition and heat treatment on the structure of cast irons. From W. G. Moffatt, G. W. Pearsall, and J. Wulff, *Structure and Properties of Metals*, v. 1, Wiley (1964).

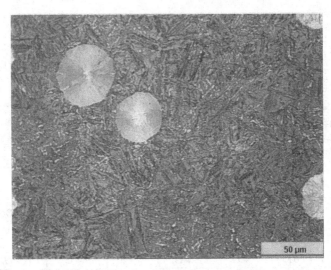

Figure 21.23. Grade 900 austempered ductile cast iron. The matrix is ausferrite consisting of 35% austenite and 65% ferrite. Courtesy of Kathy Hayrynen and John Keough, Applied Process, Inc.

wide range of uses. Austempering of ductile cast iron is quite different from that of steel. The microstructural products are an acicular ferrite that grows around the graphite nodules. As these grow, the austenite is enriched in carbon up to about 2%, which is stable. No bainite is present. The stabilized austenite will not transform to martensite at subzero temperatures but will transform on plastic straining. This strain-induced transformation generates good wear resistance. Figures 21.23 and 21.24 show the microstructure on two austempered ductile irons.

On a strength-to-weight ratio, austempered ductile iron is superior to aluminum alloys and forged and heat-treated steel. The strength depends on the austempering temperature (Figure 21.25).

A high impact resistance can be obtained by high austempering temperatures, which produces more stabilized austenite, but with a loss of strength (Figure 21.26).

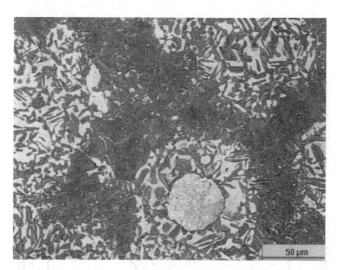

Figure 21.24. Grade 750 austempered ductile cast iron that was intercritically austeni-tized. The dark regions consist of acicular ferrite and carbon stabilized austenite. The light regions are proeutectoid ferrite. Courtesy of Kathy Hayrynen and John Keough, Applied Process, Inc.

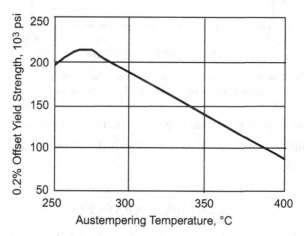

Figure 21.25. Effect of the austempering temperature on yield strength. After J. R. Keough, Applied Process Inc.

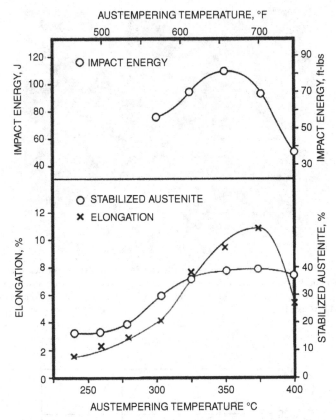

Figure 21.26. Charpy impact energy, elongation, and amount of stabilized austenite as a function of austempering temperature. From J. R. Keough, ibid.

Austempered gray iron is a material that combines the excellent damping resistance of gray iron with the strength produced by austempering.

DAMPING CAPACITY

The graphite flakes in gray cast iron give it a high damping capacity, which is useful in applications where there are mechanical vibrations. Because of its damping capacity, most machine bases are made from

Table 21.2. *Damping capacities*

Material	Damping parameter, $\partial \times 10^{-4}$*
White cast iron	2–4
Malleable cast iron	8–15
Ductile cast iron	5–20
Gray cast iron (fine flakes)	20–100
Gray cast iron (large flakes)	100–500
Steel (0.2% C)	5
Steel (eutectoid)	4

From American Cast Metals Association.
*$\partial = \ln(a_{n+1}/a_n)$, where a_{n+1}/a_n is the ratio of the amplitudes of two successive cycles.

gray cast iron. Table 21.2 compares the damping capacity of several ferrous materials.

REFERENCES

H. T. Angus, *Cast Iron: Physical and Engineering Properties* (1976).
Gray and Ductile Iron Castings Handbook, C. F. Waltron ed., Gray and Ductile Iron Founder's Society (1971).
W. F. Hosford, *Physical Metallurgy*, 2nd ed., CRC Press (2010).
A. K. Sinha, *Ferrous Physical Metallurgy*, Butterworth (1989).
Metals Handbook, 9th ed., v. 15, ASM (1988).
Typical Microstructures of Cast Iron, British Cast Iron Research Association (1951).
J. D. Verhoeven, *Steel Metallurgy for the Nonmetallurgist*, ASM (2007).

22

MAGNETIC BEHAVIOR OF IRON

GENERAL

Throughout history, magnetism has seemed a mysterious phenomenon. The discovery of lodestone (Fe_3O_4) led to many myths (Figure 22.1). Probably the first real use of the magnetic phenomenon should be attributed to the Vikings. Their development of the magnetic compass enabled them to travel far at sea even in foggy conditions. The term *magnetic behavior* usually means ferromagnetic behavior. There are actually two other types of magnetic behavior: diamagnetic behavior, which is a weak repulsion of a magnetic field, and paramagnetism, which is a weak attraction of a magnetic field.

Ferromagnetism, in contrast, is a strong attraction of a magnetic field. There are only a few ferromagnetic elements. The important ones are iron, nickel, and cobalt. A few rare earths are ferromagnetic at low temperatures. Table 22.1 lists all of the ferromagnetic elements and the temperature above which they cease to be ferromagnetic (Curie temperature).

Atoms of other transition element atoms may act ferromagnetically in alloys in which the distance between atoms is different from that in the elemental state. These include the manganese alloys Cu_2MnAl, Cu_2MnSn, Ag_5MnAl, and $MnBi$.

Figure 22.1. A typical legend about lodestone: the first permanent magnet known to man. From E. A. Nesbitt, *Ferromagnetic Domains*, Bell Telephone Labs (1962).

FERROMAGNETISM

The physical basis for ferromagnetism is an unbalance of electron spins in the 3-d shell of the transition elements and the 4-f shell for rare earths. An unbalanced spin causes a magnetic moment. In metals

Table 22.1. *Ferromagnetic elements and Curie temperatures*

Metal	Curie temperature, °C
Iron	1044
Cobalt	1121
Nickel	358
Gadolinium	16
Terbium	−40
Dysprosium	−181

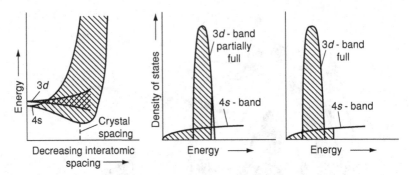

Figure 22.2. For the transition elements, the 3-d and 4-s energy levels overlap. From R. E. Smallman, *Modern Physical Metallurgy*, 2nd ed., Butterworth (1963).

with valences of 1 or 3 (e.g., Cu or Al), each atom has an unbalance of spins, but the unbalance is random, so there is no net effect. With the transition elements, the 3-d and 4-s energy bands overlap at the distance between atoms in the crystal (Figure 22.2).

For some, the total energy is lowered in a magnetic field if half of the 3-d band is completely full, causing an imbalance of electron spins, as shown schematically in Figure 22.3. This results in a strong magnetic effect. The field caused by neighboring atoms is strong enough to cause this shift.

This lowering of energy caused by alignment of the unbalanced spins with that of the neighboring atoms is called the exchange energy. It depends on the interatomic distance. For example, body-centered cubic iron is ferromagnetic but face-centered is not (Figure 22.4). Figure 22.5 shows how the maximum number of unbalanced spins per atom (number of Bohr magnetons) depends on the number of 3-d electrons.

Ferromagnetic behavior depends on four important energy terms:

1. *exchange energy*, which has already been discussed;
2. *magnetostatic energy*;
3. *magnetocrystalline energy*; and
4. *magnetostrictive energy*.

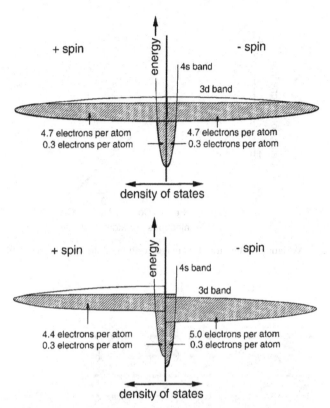

Figure 22.3. If one-half of the 3-d band is completely full and the other half partially full, there is a strong unbalance of electron spins, causing a strong magnetic effect. From W. F. Hosford, *Physical Metallurgy*, 2nd ed., CRC Press (2010).

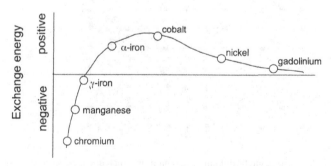

Figure 22.4. Dependence of exchange energy on atomic separation. Adapted from J. K. Stanley, *Electrical and Magnetic Properties of Metals*, ASM (1963).

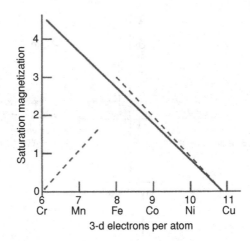

Figure 22.5. Variation of the saturation magnetization on the number of 3-d electrons.

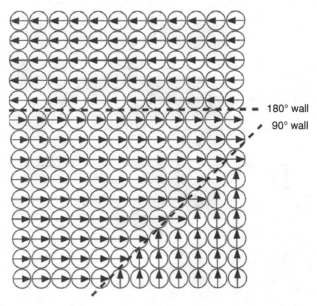

Figure 22.6. Ferromagnetic domains are regions in which unbalanced electron spins are aligned. Parts of three domains are indicated. The dashed lines are 180° and 90° domain walls.

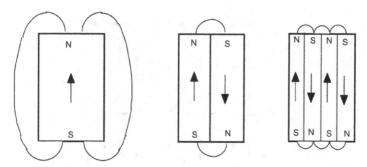

Figure 22.7. Incomplete magnetostatic circuits raise the energy.

Exchange energy is minimized if neighboring atoms are magnetized in the same direction. This causes the formation of magnetic domains in which all of the neighboring atoms are magnetized in the same direction. These may contain 10^{15} atoms. Figure 22.6 schematically shows parts of two neighboring domains.

MAGNETOSTATIC ENERGY

However, incomplete magnetostatic circuits raise the total energy because the circuits must be completed externally (Figure 22.7). The reason horseshoe magnets attract iron is to complete magnetostatic circuits in iron (Figure 22.8). A typical domain structure is composed of domains, which form complete circuits, as shown in Figure 22.9.

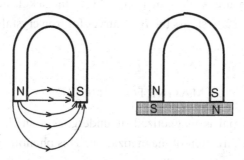

Figure 22.8. A horseshoe magnet attracts iron to complete a magnetostatic circuit.

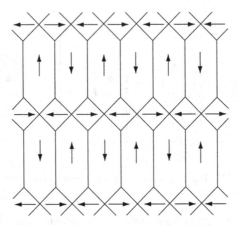

Figure 22.9. A typical domain structure composed of complete magnetostatic circuits.

When the there are equal numbers of domains aligned in opposing directions, their magnet fields cancel, and externally the material appears not to be magnetized. However, when an external magnetic field is imposed, those domains most nearly aligned with the field will grow at the expense of those aligned in the opposite direction.

MAGNETOCRYSTALLINE ENERGY

Each of the ferromagnetic materials has a specific crystallographic direction in which it is naturally magnetized. The directions of easy magnetization are <100> in iron, <111> in nickel, and [0001] in cobalt. Figure 22.10 shows the B-H curves for iron crystals of different orientations.

MAGNETOSTRICTION

When a material is magnetized, it undergoes a small dimensional change in the direction of magnetization. This phenomenon is called magnetostriction. Figure 22.11 shows the magnetostriction in the three common metals.

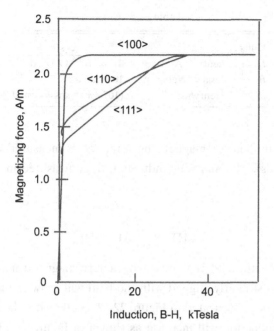

Figure 22.10. *B-H* curves for several directions in iron. Data from J. K. Stanley, ibid.

PHYSICAL UNITS

Further discussion of magnetic behavior requires definition of some terms. The units used to describe these are listed in Table 22.2, which includes both cgs and mks units and their relation. The intensity of

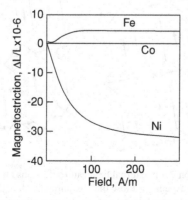

Figure 22.11. Magnetostriction of iron, cobalt, and nickel. Data from J. K. Stanley, ibid.

Table 22.2. *Units*

Units	mks	cgs
H	henry $=$ A/m	$4\pi \times 10^{-3}$ Oersted
B	tesla $=$ Weber/m^2	104 Gauss
μ_o	Henry/m	$(107/4\pi)$ Gauss/oersted

the magnetic field or magnetizing force, H, is measured in Henrys (or Oersteds). The magnetic induction, B, is measured in Teslas (or Gauss).

THE B-H CURVE

When magnetic field is imposed on a ferromagnetic material, the domains most nearly aligned with the field will grow at the expense of the others as illustrated in Figure 22.12. As they do, the material's magnetic induction will increase as shown in Figure 22.13. At first, favorably aligned domains grow. Final induction occurs by rotation of the direction of magnetization out of the easy direction to be aligned with the field. Figure 22.14 shows an entire *B-H* curve.

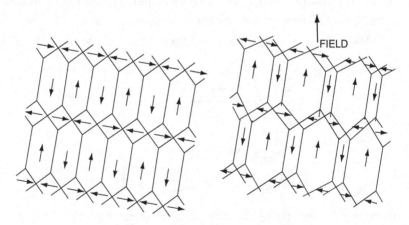

Figure 22.12. Imposition of an external field causes domains most nearly aligned with the field to grow at the expense of those that are antialigned.

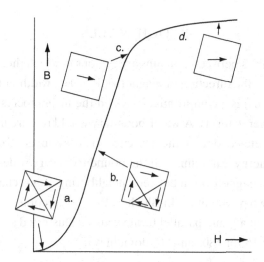

Figure 22.13. Magnetization of a material. Initially magnetization increases by growth of favorably oriented domains. At high fields, the direction of magnetization rotates out of the easy direction.

On removal of the field, there is a *residual magnetization* or *remanence*, B_r. A reverse field, H_c (*coercive force*), is required to demagnetize the material. The area enclosed by the *B-H* curve (*hysteresis*) is the energy loss per cycle. The *permeability* is defined as $\mu = B/H$. The initial permeability, μ_o, and the maximum permeability, μ_{max}, are material properties.

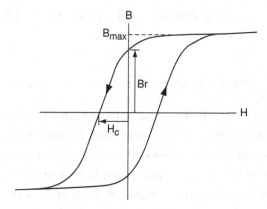

Figure 22.14. A typical *B-H* curve.

BLOCH WALLS

The boundaries between domains are regions in which there is a gradual change in the direction of magnetization. The width of these (perhaps 20 atoms) is a compromise between the magnetocrystalline and exchange energy terms. A wider boundary would require more atoms to be magnetized out of the direction of easy magnetization. The exchange energy is minimized if the boundary is very wide so that the direction of magnetization between neighboring atoms changes little. There are two possibilities. In Bloch walls, the direction of magnetization rotates in a plane parallel to the wall, as illustrated schematically in Figure 22.15 for 180° and 90° domain walls.

SOFT VERSUS HARD MAGNETIC MATERIALS

There are two main types of magnets: hard and soft. Hard magnetic materials are permanent magnets and are difficult to demagnetize. The hysteresis is very large. The remanence, B_r, and coercive force, H_c, are high. Soft magnetic materials are easily demagnetized. The terms *soft* and *hard* are historic. The best permanent magnets in the 1910s were made of martensitic steel, which is very hard, and the best soft magnets were made from pure annealed iron. The differences of the *B-H* curves are shown in Figure 22.16. Table 22.3 shows the extreme differences.

SOFT MAGNETIC MATERIALS

For a material to be soft magnetically, its domain walls must move easily. The principal obstacles to domain wall movement are inclusions and grain boundaries. Low dislocation contents and residual stresses are also important. A low interstitial content is also important.

Inclusions are important obstacles to domain wall movement because the energy of the system is lower when a domain wall passes

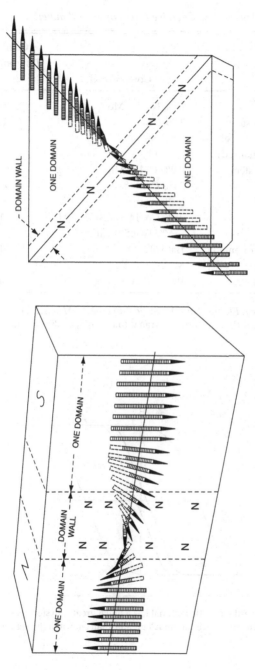

Figure 22.15. The 180° and 90° domain walls. From E. A. Nesbitt, ibid.

Table 22.3. *Coercive forces of several materials*

Material	Composition	Coercive force Hc from saturation, Oersteds
Supermalloy	79% Ni, 5% Mo	0.002
Oriented silicon steel	3.25% Si	0.1
Hot-rolled silicon steel	4.5% Si	0.5
Mild steel (normalized)*	0.2% C	4.0
Carbon magnet steel	0.9% C, 1% Mn	50
Alnico V	24% Co, 14% Ni, 8% Al, 3% Cu	600
Alnico VIII	35% Co, 14.5% Ni, 7% Al, 5% Ti, 4.5% Cu	1450
Barium ferrite (oriented)	BaO 6Fe$_2$O$_3$	1900
Bismanol	MnBi	3650
Platinum-cobalt	PtCo (77% Pt)	4300

From J. K. Stanley, *Electrical and Magnetic Properties of Metals*, ibid.
* Heating to above the transformation range followed by cooling to room temperature
  in still air.

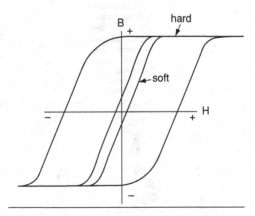

Figure 22.16. A hard magnetic material has a much greater hysteresis than a soft magnetic material. The differences are much greater than shown in this figure.

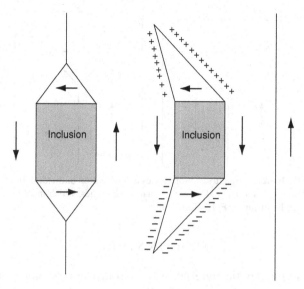

Figure 22.17. The difference between the domain boundaries at an inclusion depending on whether the inclusion lies on a boundary (left) or not (right). The total length of boundary is lowered by the inclusion. From W. F. Hosford, *Physical Metallurgy*, 2nd ed., CRC Press (2009).

through an inclusion than when the boundary has separated from the inclusion. This is illustrated in Figure 22.17.

For very soft magnetic magnets, magnetostriction should be minimized. The reason is that magnetostriction causes dimensional incompatibilities at 90° domain boundaries that must be accommodated by elastic straining of the lattic. This is illustrated in Figure 22.18.

In iron-nickel alloys, the magnetostriction and the magnetocrystalline anisotropy are low at about 78% Ni. Iron-nickel alloys have very high initial permeabilities. Mu metal (75% Ni), permalloy (79% Ni), and supermalloy (79% Ni, 4% Mo) are examples that find use in audio transformers. By heat treating in a magnetic field, a texture can be formed that has a square-loop hysteresis curve. This is useful in logic circuits where the material is magnetized in one direction or the other. For high-fidelity transformers, linearity (constant μ) is needed.

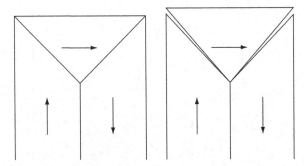

Figure 22.18. In iron, magnetostriction causes an elongation in the direction of magnetization. This creates a misfit along 90° domain boundaries, which must be accommodated elastically. From W. F. Hosford, ibid.

SILICON STEEL

Uses of soft magnetic materials include transformers, motor and generator cores, solenoids, relays, magnetic shielding, and electromagnets for handling scrap. Many of these applications employ silicon steel (usually 3 to 3.5% Si). Alloys containing 3% or more silicon are ferritic at all temperatures up to the melting point (see Figure 22.19). Silicon increases the electrical resistance of iron. A high electrical resistance is desirable for transformers because eddy currents are one of the principal power losses in transformers. Remember that power loss is inversely proportional to resistance ($P = EI = E^2/R$). The use of thin sheets also minimizes eddy current losses.

Concentrations of carbides, sulfurides, oxides, and nitrides are kept low to minimize hysteresis and prevent a loss of permeability. Carbon is especially detrimental and kept below 0.005%. There are several grades of silicon steel. Grain-oriented steel is processed to have a preferred orientation with a <100> direction aligned with the rolling direction and a {011} plane in the plane of the sheet. This texture is a result of secondary recrystallization. With this orientation (the Goss texture), the magnetic flux density is increased 30% over that of randomly oriented material. This is used in high-efficiency transformers. Because this texture has the <001> easy direction of

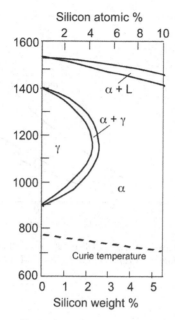

Figure 22.19. Iron-silicon phase diagram. Data from J. K. Stanley, ibid.

magnetization aligned with the prior rolling direction, transformers can be made so that they will be magnetized in a <001> direction. The cube texture, {100}<001>, is even more desirable, but it is more difficult to produce. Both the Goss and cube textures are illustrated schematically in Figure 22.20.

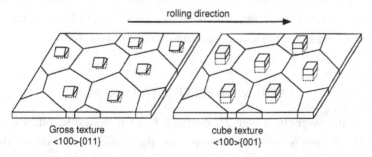

Figure 22.20. Textures in silicon. In both the Goss and cube textures, the <100> direction is aligned with the rolling direction. The {011} is parallel to the sheet in the Goss texture, and the {001} is parallel to the sheet in the cube texture. From W. F. Hosford, ibid.

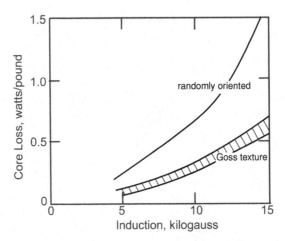

Figure 22.21. Hysteresis losses in 3.25% silicon steels at 60 cycles. Data from J. K. Stanley, *Electrical and Magnetic Properties of Metals*, ASM (1963).

Nonoriented steel has similar magnetic properties in all directions. It is less expensive than oriented steel and is used in applications such as motors and generators in which the direction of magnetization is not fixed. The core losses in transformers made from amorphous steel are about 30% of those in conventional steels, but they are much more expensive.

Core losses decrease with increasing silicon content and increase with increasing frequency. Figure 22.21 shows the core losses of transformers made with oriented and nonoriented silicon steel.

HARD MAGNETIC MATERIALS

For hard magnets, a high H_c coercive force is desirable, but a high BxH product is more important, and the second quadrant of the B-H curve (Figure 22.22) is most important. The maximum BxH product (Figure 22.23) is often taken as a beneficial feature as well.

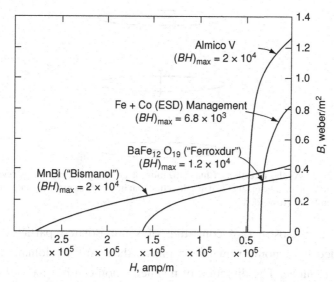

Figure 22.22. Second quadrant of *B-H* curves for selected alloys. From R. M. Rose, L. A. Shepard, and J. Wulff, *The Structure and Properties of Materials*, Wiley (1966).

A high *BxH* product results from the following:

1. small, isolated particles that are single domains;
2. elongated particles; and
3. a high magnetocrystalline energy.

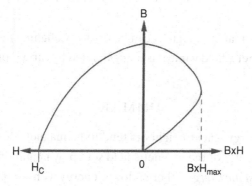

Figure 22.23. Second quadrant of a *B-H* curve (left) and the corresponding *BxH* product (right). The maximum *BxH* product is a feature of merit. From W. F. Hosford, ibid.

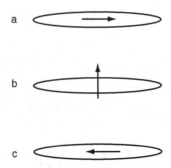

Figure 22.24. As the direction of magnetization of an elongated particle is reversed (from a to c), it must be magnetized in a direction that increases its magnetostatic energy (b). From W. F. Hosford, ibid.

In a microstructure consisting of small isolated particles surrounded by a nonferromagnetic phase, there are no domain walls that can move. The direction of magnetization can be changed only by rotating the magnetization out of one easy direction into another, equivalent easy direction. If there is high magnetocrystalline energy, it will require a high field. Hexagonal structures are useful here because there are only two easy directions, [0001] and [000$\bar{1}$], which differ by 180°. With elongated ferromagnetic particles, the intermediate stage will have a high magnetostatic energy (Figure 22.24).

Some of the best hard magnetic materials are those with a hexagonal structure, in which there are only two possible domains, differing by 180°.

Cheap permanent magnets can be made by aligning fine iron powder in a magnetic field while it is being bonded by rubber or a polymer.

SUMMARY

Exchange energy – near neighbors tend to be magnetized in the same direction because of the magnetic field set up by neighbors.

Magnetostatic energy – there is lower energy with complete magnetostatic circuits, which is why permanent magnets attract and explains the normal domain structure in soft magnetic materials.

Magnetocrystalline energy – energy is lowest when the magnetization direction is parallel to a characteristic crystalline direction. To change direction requires a high field.

Magnetostriction – the dimensional change when magnetization occurs, magnetostriction makes movement of domain walls more difficult.

REFERENCES

J. K. Stanley, *Electrical and Magnetic Properties of Metals*, ASM (1963).

E. A. Nesbitt, *Ferromagnetic Domains*, Bell Telephone Laboratories (1962).

R. M. Rose, L. A. Shepard, and J. Wulff, *The Structure and Properties of Materials, vol. IV: Electronic Properties*, Wiley (1966).

23

CORROSION

CORROSION CELLS

Corrosion is electrochemical in nature. There is always an anode, where metal goes into solution, $M \rightarrow M^{+n} + ne^-$, and a cathode where the electrons are consumed. The possible cathode reactions with iron are as follows:

$2H^+ + 2e^- \rightarrow H_2$ (This occurs only in acid solutions.)

$O_2 + 2H_2O + 4e^- \rightarrow 4(OH)^-$ (This is the most common cathode reaction. It requires oxygen dissolved in the aqueous solution.)

$O_2 + 4H^+ + 4e^- \rightarrow 2H_2O$ (Both an acid solution and dissolved oxygen are necessary.)

Table 23.1 is the electromotive series. It gives the electrode potentials of the anode half-reactions for various metals in 1 molar solutions of the metal ions. These potentials are measured against a hydrogen electrode for an anode reaction $H_2 \rightarrow 2H^+ + 2e^-$ as illustrated in Figure 23.1.

If the solution concentration is not 1 molar, the cell voltage can be found from the Nernst equation:

$$\varepsilon = \varepsilon_{25°C} + (0.00257/n)\ln C, \qquad (23.1)$$

where n is valence of the anion, and C is the molar concentration (moles per liter).

Table 23.1. *Electrode potentials (25°C, molar solutions)*

Anode half-cell reaction (the arrows are reversed for the cathode half-cell reaction)	Electrode potential used by electrochemists and corrosion engineers,* volts		
$Au \rightarrow Au^{3+} + 3e^-$	+1.50		
$2H_2O \rightarrow O_2 + 4H^+ + 4e^-$	+1.23	Cathodic	(noble)
$Pt \rightarrow Pt^{4+} + 4e^-$	+1.20		
$Ag \rightarrow Ag^+ + e^-$	+0.80		
$Fe^{2+} \rightarrow Fe^{3+} + e^-$	+0.77		
$4(OH)^- \rightarrow O_2 + 2H_2O + 4e^-$	+0.40		
$Cu \rightarrow Cu^{2+} + 2e^-$	+0.34		
$H_2 \rightarrow 2H^+ + 2e^-$	0.000	Reference	
$Pb \rightarrow Pb^{2+} + 2e^-$	−0.13		
$Sn \rightarrow Sn^{2+} + 2e^-$	−0.14		
$Ni \rightarrow Ni^{2+} + 2e^-$	−0.25		
$Fe \rightarrow Fe^{2+} + 2e^-$	−0.44		
$Cr \rightarrow Cr^{2+} + 2e^-$	−0.74	Anodic	(active)
$Zn \rightarrow Zn^{2+} + 2e^-$	−0.76		
$Al \rightarrow Al^{3+} + 3e^-$	−1.66		
$Mg \rightarrow Mg^{2+} + 2e^-$	−2.36		
$Na \rightarrow Na^+ + e^-$	−2.71		
$K \rightarrow K^+ + e^-$	−2.92		
$Li \rightarrow Li^+ + e^-$	−2.96		

* From L. H. Van Vlack, *Elements of Materials Science and Engineering*, 3rd ed., Addison-Wesley (1974).

Seawater is a common corrosive environment. The Galvanic series (Table 23.2) ranks alloys according to their electropotentials in seawater.

Corrosion cells can arise from contact of different metals. Cells can also arise within a single piece of metal between regions with different amounts of cold work, different microstructures, or different phases, as shown in Figures 23.2 and 23.3.

Variations in the concentration of oxygen in different parts of the electrolyte can set up corrosion cells. Because oxygen is needed for the cathode reaction, regions with high oxygen concentration tend to

Table 23.2. *The Galvanic series of common alloys**

Graphite	Least active
Silver	
18–8 Cr-Ni stainless steel (P)	
Nickel (P)	
Copper	
Brass	
Tin	
Lead	
Nickel (A)	
18–8 Cr-Ni stainless steel (A)	
Plain carbon steel	
Aluminum	
Zinc	
Magnesium	Most active

* P signifies the passive state, and A signifies the active state.

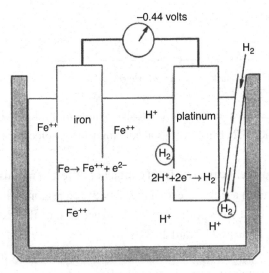

Figure 23.1. The hydrogen cathode.

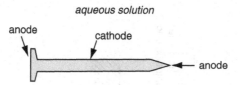

Figure 23.2. Regions that have been cold worked are anodic to regions that have not.

become cathodic. Figure 23.4 illustrates several oxygen concentration cells.

The rate of corrosion of carbon steel in water is virtually independent of its composition. It depends primarily on the access to oxygen. Figure 23.5 shows the effect of oxygen concentration in the water. Very high oxygen concentrations cause passivity.

POLARIZATION AND PASSIVITY

Current density controls the corrosion current. As corrosion occurs, the anode and cathode polarities change, as indicated in Figure 23.6. The corrosion current corresponds to the intersection of the two. It is much better to have a large anode area and a small cathode area than the reverse. A large cathode area will lead to a greater corrosion current and a greater loss of metal (see Figure 23.7). Furthermore, with the same corrosion current, the depth of corrosion on a small anode area will be greater than that on a large anode area.

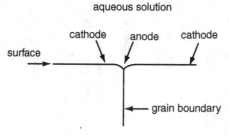

Figure 23.3. Because atoms at grain boundaries are in a higher energy state, the grain boundaries become anodic.

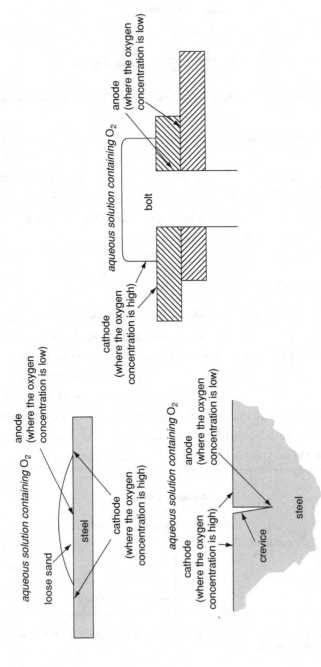

Figure 23.4. Oxygen concentration cells: the regions that are shielded from oxygen are the anodes, and the cathode reaction occurs where oxygen is plentiful.

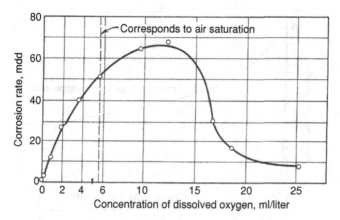

Figure 23.5. Effect of oxygen concentration on the corrosion of mild steel in slowly moving distilled water. From H. H. Uhlig, *Corrosion and Corrosion Control*, 4th ed., Wiley (2001).

The accumulation of positive ions near the anode and negative ions near the cathode polarize the cell (Figure 23.8), decreasing the voltage. Such polarization decreases the corrosion current (Figure 23.9).

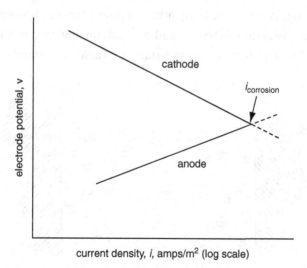

Figure 23.6. The difference between the anode and cathode potentials decreases with increasing density corrosion current.

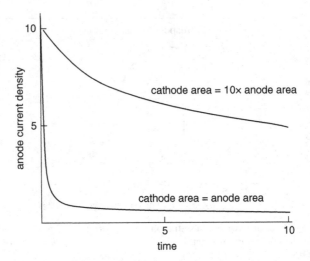

Figure 23.7. With a constant anode area, the corrosion current increases as the cathode becomes larger. The decrease of corrosion current with time is a result of polarization.

With some materials, when the anode potential reaches a critical value, the corrosion current drops abruptly to a degree that the corrosion rate is small, as shown in Figure 23.10. This condition is called passivation. A very thin oxygen layer on stainless steels is sufficient to cause passivation. Oxygen and a small amount of corrosion are required to maintain the passive state. In Table 23.2, stainless steels

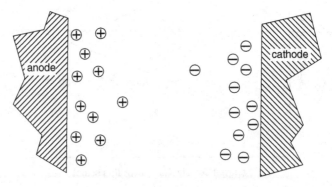

Figure 23.8. Polarization is caused by an accumulation of positive ions near the anode and negative ions near the cathode.

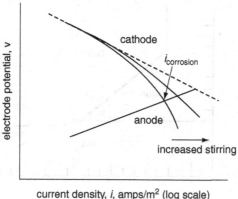

current density, i, amps/m^2 (log scale)

Figure 23.9. Polarization at the cathode decreases the cell potential. Increased convection decreases the polarization. The effects of anode polarization are similar.

occupy two places, depending on whether they are passive. Titanium alloys may be passive under special conditions.

POURBAIX DIAGRAM

Figure 23.11 is a simplified Pourbaix diagram for iron in water. It shows the effects of corrosion potential and acidity on corrosion. Since

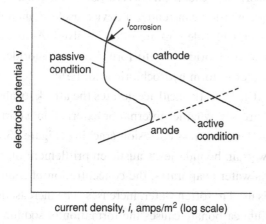

current density, i, amps/m^2 (log scale)

Figure 23.10. Above a critical anode potential, certain materials become passive. Their corrosion rate drops abruptly. Note that the current density is plotted on a logarithmic scale.

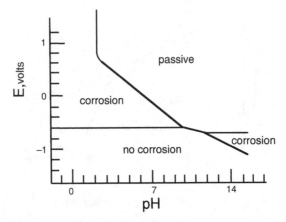

Figure 23.11. Simplified Pourbaix diagram for iron. From J. C. Scully, *The Fundamentals of Corrosion*, 3rd ed., Pergamon (1975).

it is based on thermodynamics of the iron-water system, it ignores the effects of impurities such as Cl^- and SiO_4^{2-} ions. It says nothing about kinetics and is temperature dependent.

TYPES OF CORROSION

Sometimes corrosion causes pits to form, rather than a uniform attack. Such pitting corrosion can render a device useless when 99% of the device is intact. Chloride ions are often involved. At an incipient pit, iron ions attract chloride ions to form a metal chloride, which can react with water to form hydrochloric acid. $Fe^{++} + 2Cl^- + 2H_2O \rightarrow FeOH + 2HCl$. Thus, the acid accelerates the attack locally.

Tensile stresses, whether internal or external, in a corrosive environment can lead to stress corrosion cracking (Figure 23.12). Cracks often follow grain boundaries. Caustic embrittlement of boilers is an example. As water evaporates, the concentration of sodium carbonate, which is used to soften water, increases. The increased concentration of sodium carbonate causes the formation of sodium hydroxide, $Na_2CO_3 + H_2O \rightarrow 2NaOH + CO_2$. As water evaporates, the concentration increases. This sodium hydroxide attacks the surrounding

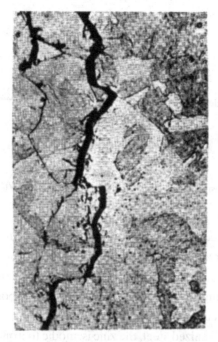

Figure 23.12. Typical stress corrosion crack in an 18% nickel maraging steel. Note how the crack follows grain boundaries. From H. L. Logan, *The Stress Corrosion Cracking of Metals*, Wiley (1966).

material and dissolves the boiler's iron, causing embrittlement. Other environments that can cause stress corrosion cracking of steels are nitrates at elevated temperatures and hydrogen sulfide, the latter of which is often encountered in drilling oil wells. It has been shown that the rate of crack growth is proportional to the current density.

CORROSION CONTROL

The means of controlling corrosion involve disrupting or changing the corrosion cell. One such method is to remove the electrolyte (corrosion will not occur in the absence of water). Corrosion can be completely prevented in closed systems by removing oxygen from the

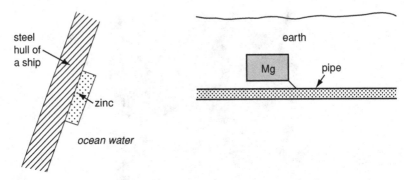

Figure 23.13. Corrosion protection by sacrificial corrosion of zinc (left) and magnesium (right).

solution. Inert coatings such as paint serve this function. Another method is to break circuit. Cells caused by electrical contact of dissimilar metals can be interrupted by placing an insulator between the metals. Reversing the voltage by imposing the opposite voltage on the external circuit and sacrificial corrosion (Figure 23.13) are other means. With galvanized steel, the zinc is anodic to iron and hence protects the steel through sacrificial corrosion rather than inert coatings (Figure 23.14).

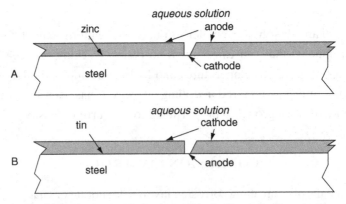

Figure 23.14. (a) Plating steel with zinc (galvanizing) offers cathodic protection to steel if the plating is scratched. (b) Tin plating offers no cathodic protection, so corrosion of the steel will occur if the plating is scratched.

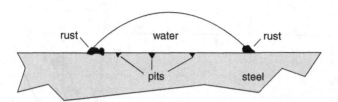

Figure 23.15. Rust formation away from corroded sites where the oxygen concentration is higher.

There are certain chemical compounds, called inhibitors, that can be used to decrease corrosion by affecting either the anode or cathode reaction. Chromates are an example of an anodic inhibitor that form a passive layer on steel. However, chromates can be cancer producing and their use is limited by law. Nitrates and phosphates also act as anodic inhibitors. Use of too little of an anodic inhibitor can lead to pitting corrosion.

RUST

Rust is ferric oxide, Fe_2O_3, and ferric hydroxide, $Fe(OH)_3$. Ferrous ions, Fe^{2+}, are soluble, but further oxidation produces ferric ions, $3Fe^{2+} + 6OH^- \rightarrow 3Fe(OH)_3 + 3H_2O$. The ferric hydroxide is insoluble, which precipitates. If dried, the ferric hydroxide turns to an oxide, $2Fe(OH)_3 \rightarrow Fe_2O_3 + 3H_2O$. Often the rust-producing reaction occurs at some distance from where the anode reaction occurs, so rust deposits may not be directly over the corroded region. This is illustrated in Figure 23.15.

DIRECT OXIDATION

It might seem as though direct oxidation in air at high temperature would not involve an electrolytic cell. However, there is an anode and cathode. The anode reaction is

$$M \rightarrow M^{n+} + ne^-$$

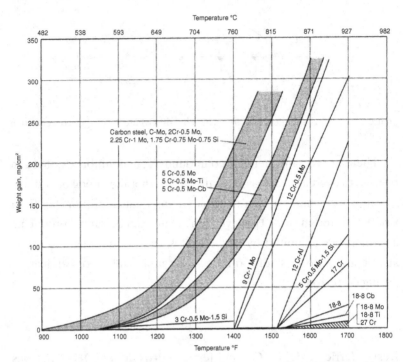

Figure 23.16. Oxidation rates of various steels. Note how the oxidation resisitence increases with chromium content. From George Y. Lai, *High-Temperature Corrosion Resistance of Engineering Alloys*, ASM (1990).

and the cathode reaction is $O_2 + 4e^- \rightarrow 2O^{2-}$. Either O^{2-} ions or M^{n+} ions and e^- must diffuse through the oxide. M^{n+} ions are smaller than O^{2-} ions and therefore diffuse faster; hence their diffusion is rate controlling. (This was illustrated in Figure 19.12.)

Cr_2O_3 is generally defect free, so ion diffusion and electron transport are slow. Hence, steels containing chromium have increased resistance to oxidation. This is evident in Figure 23.16.

With iron, several oxide layers develop, depending on temperature (Figure 23.17).

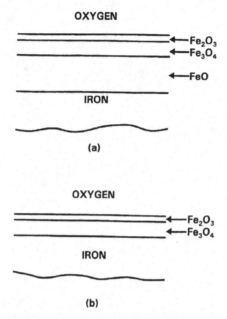

Figure 23.17. Sequence of oxides above 570°C (a) and below 570°C (b). From D. Talbot and J. Talbot, *Corrosion of Iron and Steel*, CRC Press (1998).

REFERENCES

ASM Metals Handbook, v. 13: Corrosion, 9th ed. (1987).

J. C. Scully, *The Fundamentals of Corrosion*, 3rd ed., Pergamon Press (1990).

H. H. Uhlig, *Corrosion and Corrosion Control*, Wiley (1963).

L. L. Shreir, R. A. Jarman, and G. T. Burstein, eds., *Corrosion: Metal/ Environmental Reactions*, 3rd ed., Butterworth Heineman (1994).

D. Talbot and J. Talbot, *Corrosion Science and Technology*, CRC Press (1998).

The Making, Shaping and Treating of Steel, 9th ed., U.S. Steel Corp. (1971).

PHYSICAL PROPERTIES
OF PURE IRON

Atomic number	26		
Atomic weight	55.845 amu		
Isotopes	52	Half-life	8.3 hrs
	54	Stable	
	55	Half-life	2.7 yrs
	56	Stable	
	57	Stable	
	58	Stable	
	59	Half-life	54.5 days
	60	Half-life	1.5×10^6 yrs
Atomic radius	0.1241 nm		
Energy of a vacancy (bcc)	6.006 eV		
Energy of a vacancy (fcc)	5.858 eV		
Phases and phase change temperatures			
	Melting point	2781°C	
	Boiling point	$3000 \pm 150°C$	
	α-γ transformation	$910 \pm 1°C$	
	γ-α transformation	$1390 \pm 5°C$	
Latent heats			
Latent heat of fusion	65.5 cal/g		
	α-γ transformation	3.85 cal/g	
	γ-α	2.95 cal/g	
Density	0.874 Mg/m³ at 20°C		
Thermal expansion 20 to 100°C	$11.55 \times 10^{-6}/°C$		
Specific heat	0.106 cal/°C-mole		
Thermal conductivity	0.2 cal/(cm·sec·°C)		

(continued)

Electrical resistivity	9.71×10^{-6} ohm-cm
Temperature coefficient of resistivity	0.651% increase/$^\circ$C
Saturation magnetic moment	1.714 per cm^3
Saturation magnetostriction	20×10^{-6}
Crystalline anisotropy	0.0460 J/cm^3
Curie temperature	775°C
Elastic constants	
s_{11}	5.27×10^{-8} psi
s_{12}	-1.99×10^{-8} psi
s_{44}	6.17×10^{-8} psi
c_{11}	34.9×10^{-6} (psi)$^{-1}$
c_{12}	21.2×10^{-6} (psi)$^{-1}$
c_{44}	16.2×10^{-6} (psi)$^{-1}$
	200
Young's modulus	28.9×10^{6} psi
Shear modulus	11.64×10^{6} psi
E_{111}/E_{100}	2.16
Slip systems	<111> direction {110}, {112}, and {123} planes, pencil glide
Twining systems	<111> direction {112} plane
Cleavage plane	{100}
Free surface energy	2500 mJ/m^3
Liquid-solid interfacial energy	20.8 mJ/m^3
High angle grain boundary (γ-iron)	756 mJ/m^3
High angle grain boundary (Fe-3 wt % Si)	317 mJ/m^3
	304 Stainless steel 835 mJ/m^3
Coherent twin boundary (304 stainless steel)	19 mJ/m^3
Incoherent twin boundary (304 stainless steel) 209 mJ/m^3	

201

APPENDIX II

APPROXIMATE HARDNESS CONVERSIONS AND TENSILE STRENGTHS OF STEELS

Hardness	Tensile strength			
RC	RA	V RB	B	(MPa)
68	85.6	940		
67	85	900		
66	84.5	865		
65	83.9	832	(739)	
64	83.4	800	(722)	
63	82.8	772	(705)	
62	82.3	746	(688)	
61	81.8	720	(670)	
60	81.2	697	(654)	
59	80.7	674	(634)	
58	80.1	653	615	
57	79.6	633	595	
56	79.0	613	577	
55	78.5	595	570	2075
54	78.0	577	543	2013
53	77.4	560	525	1951
52	76.8	544	512	1882
51	76.3	528	496	1820
50	75.9	513	481	1758
49	75.2	498	469	1696
48	74.7	484	455	1634
47	74.1	471	443	1579
46	73.6	458	432	1531
45	73.1	446	421	1482

(*continued*)

Hardness					Tensile strength
44	72.5	434		409	1434
43	72.0	423		400	1386
42	71.5	412		390	1338
41	70.9	402		381	1296
40	70.4	392		371	1248
39	69.9	382		362	1213
38	69.4	372		353	1179
37	68.9	363		344	1158
36	68.4	354	(109)	336	1117
35	67.9	345	(108.5)	327	1082
34	67.4	336	(108)	319	1055
33	66.8	327	(107.5)	311	1027
32	66.3	318	(107)	301	1000
31	65.8	310	(106)	294	979
30	65.3	302	(105.6)	286	951
29	64.7	294	(104.5)	279	931
28	64.3	286	(104)	272	910
27	63.8	279	(103)	266	883
26	62.8	272	(102.5)	258	862
25	62.5	266	(101.5)	253	841
24	62.0	260	(101)	247	827
23	61.5	254	100	243	807
22	61.0	248	99	237	786
21	60.5	243	98.5	231	772
20		238	97.8	226	758
(18)		230	96.8	219	731
(16)		222	95.5	212	703
(14)		213	93.9	203	676
(12)		204	92.3	194	648
(10)		196	90.7	187	621
(8)		188	89.5	179	600
(6)		180	85.5	171	579
(4)		173	84.5	165	552
(2)		166	83.5	158	531
(0)		160	81.8	152	517

RC   (Rockwell C scale)
RA   (Rockwell A scale)
V (Vickers
  hardness)
RB   (Rockwell B scale)
B   (Brinell hardness – 3000-kg
              load)

INDEX

AISI designation system, 195
alloy steels, 195
alloying element effects, 195
aluminum killing, 108, 167
anisotropy, 180
anisotropic yielding, 183–186
annealing, 51–64
Arrhenius equation, 56, 96
atomic diameters, 91
austempered ductile cast iron, 250
austempering, 135
austenite, 25
austenite formation, 113
austenitic stainless steel, 209
Avrami, 56

bainite, 123
bainite formation, 195
bake hardening, 108, 172
banding, 28
basic oxygen process, 16
Bessemer process, 13
beta iron, 2
blast furnace, 3
B-H curve, 264
Bloch walls, 266
boron, 146
brittle fracture, 222

carbide formers, 88
carbides in tool steels, 20
carbon equivalent, 243
carbon solubility, 94
carboaustenitizing, 163

carbonitriding, 164
carburizing, 158
case hardening, 165
cast iron, 234–255
cast iron matrices, 246
casting, 20
Catalan furnace, 6
cementite, 25
Charpy test, 226
chromium equivalents, 211
cleavage, 222
coke, 13
cold rolling, 22
combined temperature and strain rate
    effects, 46
compact graphite iron, 242
complex phase steels, 173
compression texture, 72
continuous cooling diagrams, 123
corrosion, 276–289
corrosion cells, 276–289
corten, 174
critical diameter, 142
crucible steel, 9
crystallographic textures, 183
cup-and-cone fracture, 218
Curie temperature, 256
curly grains, 67

Damascus steel, 8
damping, 254
decarburizing kinetics, 162
deep drawing, 187
deoxidation, 19

diffuse necking, 191
diffusion, 98–103
   mechanisms, 99
discontinuous grain growth,
   63
dislocation density, 35
domain wall movement, 264
domain walls, 261
drawn wires, strength, 172
dual-phase steels, 172
ductile cast iron, 244
ductile fracture, 218
duplex stainless steels, 213
dynamic strain aging, 109

electric arc process, 10
electrode potential, 276
eutectoid composition, 86
eutectoid temperature, 83
eutectoid transformation, 31, 80–81
exaggerated grain growth, 63
exchange energy, 258

fatigue, 229
ferrite, 2, 25
ferritic stainless steel, 205
ferromagnetism, 250
flow stress
   grain size effect, 37
   solute effect, 35
   strain-rate effect, 40
   temperature effect, 39
forming limits, 191
fracture, 218–232
fracture types, 218
full anneal, 26
furnace atmospheres, 165
furnace linings, 19

galvaneal, 177
galvanic series, 22
galvanizing, 22
Goss texture, 271
grain boundary migration, 61
grain growth, 58
grain size, 37
graphite flakes, 237
gray cast iron, 237
grain boundary migration, 61

Hadfield manganese steel, 198
Hall-Petch, 37
hard vs. soft magnets, 266
hardenability, 137–148
hardenability bands, 140
hardness, 293
heating during deformation,
   175
hot rolling, 21
HSLA steels, 172
Hultgren extrapolation, 33
hydrogen embrittlement, 227
hydrostatic pressure, 142
hypereutectoid, 25
hypoeutectoid, 25
hysteresis, 265

ideal diameter, 112
ideal quench, 112
inclusion-shape control, 178
inherently fine grain size, 63
intergranular fracture, 224
interstitial diffusion, 90, 102
interstitial lattice expansion, 92
interstitial solute effects, 87
interstitial-free steel, 169
iron ores, 11
isothermal transformation, 120

Jominy test, 137

killing, 20, 171
kinemativ hardening, 187

limiting drawing ratio, 189
liquid metal embrittlement, 227
localized necking, 190
low-carbon steel grades, 167–178
Lüder's bands, 104

magnesium addition, 16, 244
magnetic behavior, 256–273
magnetic domains, 261
magnetic units, 263
magnetocrystaline energy, 262
magnetostatic energy, 261
magnetostriction, 262
malleable cast iron, 244
manganese sulfide, 30, 178, 221

maraging steel, 198
marquenching, 133
martensite, 125
    lath, 86
    twinned, 86
martensite transformation, 128
martensite types, 132
martensitic sheet steel, 173
martensitic stainless steel, 208
mechanical fibering, 183
meteorite, 4
microalloying, 172
multiplying factors, 144

native iron, 4
Nernst equation, 276
nitriding, 163
nitrogen solubility, 94
nodularity in ductile iron, 244
normalizing, 26

open-hearth process, 15
orange peel, 179
oxidation resistance, 216, 287
oxygen concentration cells,
    277

partitioning of solutes, 115
passivity, 279
pearlite, 25
pearlite formation, 30, 113
pearlite spacing, 32
permeability, 265
pelletizing, 11
Persian sword, 8
phase diagrams, 54
    Fe-C, 93, 234
    Fe-Cr, 83, 206
    Fe-Cr-C, 210
    Fe-Cr-Ni, 212
    Fe-Mn, 82
    Fe-Mo, 83
    Fe-Mo-C, 85
    Fe-W-C, 85
pig iron, 12
pit furnaces, 4
polarization, 279
polygonization, 53
Pourbaix diagram, 283

precipitation-hardening stainless steels,
    212
preferred orientation, 183
proeutectoid, 25

quench severity, 143

recovery, 51
recrystalization, 54
recrystalization texture, 77
recycling, 22
residual magnetization, 265
residual stress, 54
retained austenite, 126
ridging, 126
rimming, 20, 171
roller leveling, 107
rolling texture, 71
roping, 207
rust, 287
R-value, 180

sandwich sheets, 176
secondary hardening, 156
secondary recrystalization, 63–64, 270
sensitization, 5
shaft furnace, 4
sheet forming, 179–193
sheet steel, 167–178
sigma phase, 207
silicon steel, 270
slags, 19
slip systems, 66
Snoek effect, 95
solid solution hardening, 88
solute effect on tensile strength,
    39
solute segregation, 115
spheroidite, 89, 154
spheroidizing, 27
stainless steels, 205–217
strain aging, 104–112
strain hardening, 35, 179
strain-rate effects, 40, 180
static fatigue, 228
strength differential effect, 49
strength of low-carbon steel, 167
stress relief, 54
stretcher strains, 179

stretching, 187
subgrains, 53, 73
superplasticity, 49
surface finish, 179
surface treatments, 177
swaged wires, 70

Taylor-welded blanks, 175
temperature dependence of flow stress,
    39
temper embrittlement, 157
temper rolling, 107
tempering, 150–157
texture formation, 72
tool steel heat treatment, 201
tool steels, 199
tramp elements, 226
transition temperature, 24
TRIP steels, 226

tundish, 20
tuyeres, 11
twinning, 66

vermicular graphite, 242
void coalescence, 218

weathering steel, 174
white cast iron, 236
Widmannstätten, 4, 26
wire textures, 66
wootz, 8
wrinkling, 190
wrought iron, 6

yielding, 104
Young's modulus, 87

Zener-Holomon, 104